Nature's Cornucopia: Our Stake in Plant Diversity

WITHDRAWN

JOHN TUXILL

Jane A. Peterson, *Editor*

WORLDWATCH PAPER 148

September 1999

THE WORLDWATCH INSTITUTE is an independent, nonprofit environmental research organization in Washington, DC. Its mission is to foster a sustainable society in which human needs are met in ways that do not threaten the health of the natural environment or future generations. To this end, the Institute conducts interdisciplinary research on emerging global issues, the results of which are published and disseminated to decision-makers and the media.

FINANCIAL SUPPORT for the Institute is provided by the Geraldine R. Dodge Foundation, the Ford Foundation, the William and Flora Hewlett Foundation, W. Alton Jones Foundation, Charles Stewart Mott Foundation, the Curtis and Edith Munson Foundation, David and Lucile Packard Foundation, V. Kann Rasmussen Foundation, Rockefeller Financial Services, Summit Foundation, Turner Foundation, U.N. Population Fund, Wallace Genetic Foundation, Wallace Global Fund, Weeden Foundation, and the Winslow Foundation. The Institute also receives financial support from its Council of Sponsors members—Tom and Cathy Crain, Toshishige Kurosawa, Kazuhiko Nishi, Roger and Vicki Sant, Robert Wallace, and Eckart Wintzen—and from the Friends of Worldwatch.

THE WORLDWATCH PAPERS provide in-depth, quantitative and qualitative analysis of the major issues affecting prospects for a sustainable society. The Papers are written by members of the Worldwatch Institute research staff and reviewed by experts in the field. Published in five languages, they have been used as concise and authoritative references by governments, nongovernmental organizations, and educational institutions worldwide. For a partial list of available Papers, see back pages.

Table of Contents

The views expressed are those of the author and do not necessarily represent those of the Worldwatch Institute; of its directors, officers, or staff; or of its funding organizations.

ACKNOWLEDGMENTS: I thank Don Duvick, Alan Hamilton, Brien Meilleur, Margery Oldfield, and Sandra Postel for sharing their knowledge and providing extremely helpful comments and suggestions on drafts of this paper. At Worldwatch, I am indebted as well to Janet Abramovitz, Chris Bright, Lester Brown, Chris Flavin, Brian Halweil, and Ashley Mattoon for reviewing the paper from start to finish and helping track down information sources. As editor, Jane Peterson provided a steady hand and keen eye for clarity in polishing the manuscript. Payal Sampat reviewed the references and, along with Hilary French, clarified key points in the text. Mary Caron, Dick Bell, Amy Warehime, and Alison Trice did a masterful job on production and outreach; and Liz Doherty serenely performed her usual typesetting magic. Special thanks to Julie Velasquez Runk for her advice and support in all stages of this project.

JOHN TUXILL is a Research fellow with the Worldwatch Institute, where he researches and writes about the conservation of biological diversity. He is a co-author of the Institute's 1998 and 1999 reports *State of the World* and also wrote Worldwatch Paper 141, on the implications of declines in vertebrate species. Currently he is pursuing a doctoral degree in the School of Forestry and Environmental Studies at Yale University. He also holds a master's degree in Conservation Biology and Sustainable Development from the University of Wisconsin and a bachelor's degree from Williams College. He has conducted fieldwork in Panama, Costa Rica, Dominica, and the southwestern United States.

Introduction

L ooking back, the 1990s may be remembered as the decade
that biotechnology began to bloom—at least that is what
its boosters in agriculture and government hope. They her-
ald genetic engineering as not only a dazzling new way of
breeding crops, but also a key to achieving food security and
sustainable agriculture in the century ahead. Since 1994,
biotech companies have introduced more than 50 genetically engineered or "transgenic" varieties of crops, so named
because their DNA contains genes transferred from other
organisms that provide resistance to herbicides, pests, and
diseases. And commercial farmers have adopted transgenic
crops at a torrid pace. In 1999, farmers will plant them on
more than 50 million hectares worldwide, three times more
than the previous year.[1]

One of the latest transgenic crops is a line of potato
varieties called "NewLeaf®," developed by the Monsanto
Corporation. The DNA of these plants contains an exotic
gene from a soil bacterium that codes for the production of
a potent natural insecticide called Bt. As each NewLeaf potato plant grows, it generates Bt throughout its tissues—in
stems, leaves, roots, flowers, and tubers. Thus armed against
insects like the Colorado potato beetle, NewLeaf potatoes
need fewer applications of chemical insecticides—an important advancement given that by harvest time conventional
potatoes have, on average, been doused in more pesticides
than any other major food crop.[2]

Yet the biggest problem for potato farmers worldwide is
not beetles but a fungal disease called late blight. It was late

blight that colonized and devastated the genetically uniform potato fields of Ireland in the 1840s, triggering the famine that claimed more than a million lives. The disease, which is native to Mexico and spread worldwide in the 1800s, has been controlled in the twentieth century primarily with fungicides, but in the mid-1980s farmers began reporting outbreaks of fungicide-resistant blight. These newly virulent strains have cut global potato harvests in the 1990s by 15 percent, a $3.25 billion yield loss; in some regions, such as the highlands of Tanzania, losses to late blight have approached 100 percent. Commercial potato growers in the United States must apply as many as eight doses of powerful fungicides each growing season to stave off blight in their fields.[3]

To combat late blight, agricultural research institutions like the International Potato Center (CIP) in Lima, Peru, have turned not to Monsanto's laboratories, but to Quechua and Aymara peasant farmers high in the Andes mountains of Peru, Ecuador, and Bolivia. These peoples' ancestors first domesticated potatoes roughly 7,000 years ago, and today indigenous Andean communities still cultivate the world's richest array of potato varieties—sometimes as many as a hundred kinds in a single village. Over the past three decades, CIP researchers have collected more than 3,000 potato varieties from Andean farmers' fields, along with the scraggly wild relatives of potato plants that often grow nearby. Plant breeders the world over draw on the unique genetic traits of these traditional potatoes to breed new varieties. Without the collections at CIP and other "gene banks," it is doubtful that potatoes could continue to be grown commercially. Nor are they an exception among crops; plant breeders annually draw on traditional varieties and wild relatives for about 6 percent of the germplasm lines used in breeding new crop varieties.[4]

Although we have achieved unprecedented skill in moving genes around, only nature can manufacture them. For all its appeal, agricultural biotechnology relies on the same resource that traditional indigenous farmers do: biological diversity, or "biodiversity" for short—a word that refers to the richness and complexity of life on Earth. Biodiversity is most commonly measured as numbers of species, but its full scope is much greater. It also encompasses the different genes that individual organisms carry; the distinct populations, varieties, and breeds evident within species; and the patchwork of natural communities that species assemble into when they share a common habitat—all part of the kaleidoscopic variation of nature produced by 3 billion years of evolution.[5]

The future of every human activity that utilizes biological resources depends on our stewardship of biodiversity. For this reason, the disquieting trends now evident in Earth's natural systems should give us pause. The scale of human enterprise has become so great—we are now 6 billion strong and consume about 40 percent of the planet's annual biological productivity—that we are altering nature's ecological foundations dramatically. Current rates of species extinction are now thought to be at least 100 to 1,000 times higher than normal or "background" levels. More than 20,000 species of plants, birds, mammals, molluscs, crustaceans, and other organisms are declining in numbers or persist only in small isolated populations that place their future in doubt. With the biological scaffolding of our planet simplified across increasingly large areas, entire natural communities are now in danger of disappearing.[6]

One group of organisms—green plants—shows just how much is at stake when we erode the variation of life. No subset of biodiversity is so closely linked to humankind's past and future as the plant kingdom; indeed, a substantial chunk of plant diversity—the hundreds of thousands of varieties of cultivated plants—is the direct result of our own labor and ingenuity. In addition to providing the genetic underpinnings of our food supply, plant diversity keeps us healthy—one in every four drugs prescribed in the United States is based on a chemical compound originally discovered in plants. Plants also furnish oils, latexes, gums, fibers, timbers, dyes, essences, and other products that we use to clean, clothe, shelter, and refresh ourselves and that have many industrial uses as well. Healthy assemblages of native plants renew and enrich soils, regulate our freshwater supplies, prevent soil erosion, and provide the habitat needed by animals and other creatures. Indeed, biologists think that one reason why insects are by far the most diverse animal group is that they have evolved in close concert with plants, as both pollinators and herbivores.[7]

It is not just obscure or seemingly unimportant plants that are in trouble—those that we rely upon most heavily are declining too. Traditional crop varieties are falling by the wayside as farmers change agricultural practices and, in particular, move toward growing a less diverse pool of varieties favored for commercial production. In the United States, only 5 to 20 percent (depending on the crop) of varieties registered in a 1904 inventory are still grown commercially or stored in the national plant germplasm collection system. China is thought to have lost nearly 90 percent of its traditional wheat varieties since World War II. Many wild medicinal plants experiencing heavy commercial demand are harvested unsustainably, making them increasingly hard to obtain. Habitat loss, too, is shrinking the populations of many useful plants.[8]

Our increasingly global way of life also has unintended negative consequences for the health of the plant world. As we move more frequently about the planet, other species move too—such as weed seeds in a shipment of grain, or ornamental plants for landscaping. Too many of these translocated species escape into new habitats and become unwanted *invasives*, outcompeting native species and triggering ecological disruptions. The greenhouse gases we emit into the Earth's atmosphere in increasing volumes are also leading to unexpected changes in plant communities, many of which are exquisitely sensitive to shifts in climate and carbon dioxide concentrations.[9]

The presence of such unintended consequences means we must manage our use of plant diversity carefully. Consider the potential side effects of genetically engineered crops. Laboratory studies have shown that pollen from Bt-engineered corn is toxic to the larvae of some non-pest insects, such as the monarch butterfly, famed for its continent-wide migrations. In addition, crops that constantly manufacture their own Bt toxin increase the odds that pests like potato beetles and corn borers will develop resistance to the toxin, a worrisome prospect for those farmers who have used Bt directly for decades as a natural, relatively selective chemical control. In certain situations, transgenic crops also could transfer their genes for herbicide resistance into nearby populations of weedy wild relatives with whom they interbreed, creating the possibility of new "superweeds." European Union environment ministers, reflecting broad public sentiment across Europe, recently decided they need to know more about these risks, and suspended all licensing of new transgenic crops for planting or import into Europe.[10]

To be sure, our development path this century has yielded benefits: we now grow more food than ever before, and those who can purchase it have a material standard of living unimaginable to earlier generations. But with the world projected to add several more billion people in the next 50 years, we have to continue raising agricultural yields without dramatically expanding the total area we cultivate or the total volume of water our crops consume. Public health programs will continue to need new drugs to ward off emerging health concerns such as rising antibiotic resistance and resurgent tuberculosis and malaria. Governments still need options for raising the living standards of the 1.3 billion people who live in abject poverty and rely on plant resources for many subsistence needs. If we continue to erode the diversity of plants, these and other advancements become all the more difficult to achieve and sustain.[11]

Maintaining plant diversity begins with conserving what remains, in gene banks, botanical gardens, and national parks and protected areas systems. An even more daunting challenge, however, is revitalizing cultural practices that foster plant diversity, and reforming those that work against it. Our predominant systems for producing food, for instance, favor centralization and economies of scale, which discourage variation and local adaptation of the gene pools of crops and the ecological mosaics covering farm landscapes. And while some international bodies like the Convention on Biological Diversity (CBD) require governments to develop policies for managing biological resources wisely, others like the World Trade Organization (WTO) demand that countries dismantle them—as barriers to free trade.

Changing such broad social patterns begins by raising peoples' awareness of the benefits plant diversity provides and the threats it faces. On this point there is reason for hope: from farmers to biochemists to bankers, people increasingly view biodiversity as a resource whose sustainable use makes sound economic as well as ecological sense. Of course, putting biodiversity to use in ways that do not degrade it is nothing new. A vast pool of practical knowledge about plants, animals, and ecological patterns exists within the traditions, languages, stories, and oral histories of indigenous and peasant cultures worldwide—but this "ethno-ecological" wisdom is vanishing even more rapidly than nature itself, as rural societies face accelerating social change. Revitalizing time-tested, hands-on ways of learning about the places we inhabit may prove just as important for our future as the new ways of computers and the Internet. And the first step in finding a common way forward is to learn just how much we have to lose.[12]

Bio-uniformity Rising

Since the 18th century, when the Swedish botanist Linnaeus developed a modern scientific system for classifying all forms of life, scientists have named and described more than 275,000 species of "higher" plants, a group that includes flowering plants, conifers, ferns, and horsetails. (Their evolutionary kin, the "lower" plants, are algae, seaweeds, and mosses.) Scientists also estimate there may be upwards of 50,000 higher plant species yet to be documented, primarily in the remote, little-studied reaches of tropical forests.[13]

Although biological fluctuations are part of the ebb and flow of evolution, the loss of species and other large swaths of the ecological web are normally rare events. The natural, or "background," rate of extinction as calculated from the fossil record of life on Earth appears to be on the order of one to 10 species a year. By contrast, scientists estimate that extinction rates have accelerated this century to at least 1,000 species per year. These numbers indicate we now live in a time of mass extinction—a global evolutionary upheaval in the diversity and composition of life on Earth.[14]

Paleontologists studying Earth's fossil record have identified at least five previous mass extinction episodes during life's 3.2 billion years of evolution, with the most recent occurring about 65 million years ago, at the end of the Cretaceous period, when the dinosaurs disappeared. Earlier mass extinctions hit marine invertebrates and other animal groups hard, but plants weathered most of these episodes with relatively little trouble. Flowering plants, which now comprise nearly 90 percent of all land plant species, do not even appear in the fossil record until the Cretaceous—a relatively recent emergence in evolutionary terms.[15]

In our current mass extinction, however, plants are suffering unprecedented losses. According to a 1997 global *Red List* analysis of more than 240,000 higher plant species, which was coordinated by the World Conservation Union (IUCN), one out of every eight plants surveyed is potentially at risk. (See Table 1.) This tally includes almost 15,000 species known to be endangered or clearly vulnerable to extinction. Some endangered species are right on the edge. *Lotus kunkelii*, for one, a kind of clover native to the Canary Islands, is now reduced to a single population covering less than 500 square meters. The tally also includes about 14,500 species that are naturally rare (and thus at risk from ecological disruption), and 4,000 of indeterminate status—so poorly known that they could be dwindling without our even realizing it.[16]

Both the IUCN *Red List* and individual country rosters

TABLE 1

Threatened Plant Species, 1997

Status	Total	Share
	(number)	(percent)
Total Number of Species Surveyed	242,013	
Total Number of Threatened Species	33,418	14
Vulnerable to extinction	7,951	3
In immediate danger of extinction	6,893	3
Naturally rare	14,505	6
Indeterminate status	4,070	2
Total Number of Extinct Species	380	<1

Source: See endnote 16.

of endangered plants are dominated by *endemic* species—those found only in a relatively restricted area, such as a single country or state, an isolated mountain range, or a specific soil type. Nutrient-poor soils with high concentrations of the mineral serpentine, for instance, are famous for hosting a distinctive endemic flora wherever they occur. More than 90 percent of the "at-risk" species on the *Red List* are endemic to a single country—that is, found only in that country and nowhere else in the world. Isolated island groups are particularly rich in endemic species, nowhere more so than the Hawaiian Islands, where 91 percent of the native species are endemic. Due to their long evolutionary isolation and naturally small ranges, endemic island species are particularly vulnerable to competition from non-native species and to loss of habitat. Presently, some 20 percent of native Hawaiian plant species are candidates for the U.S. endangered species list. A dozen of these are now down to only a single wild individual each.[17]

The United States, Australia, and South Africa have the most plant species at risk of extinction. (See Table 2.) The high standing of the first two countries, however, is due in large part to how much better known their flora is compared

TABLE 2

Top 10 Countries with the Largest Numbers of Threatened Plants

Country	Total Species	Percentage of Country's Total Flora Threatened
	(number)	
United States	4,669	29
Australia	2,245	14
South Africa	2,215	11.5
Turkey	1,876	22
Mexico	1,593	6
Brazil	1,358	2.5
Panama	1,302	13
India	1,236	8
Spain	985	19.5
Peru	906	5

Source: See endnote 18.

with other species-rich countries such as Brazil and Nigeria. We have a good idea of how many plants have become endangered as the coastal sage scrub and perennial grasslands of California have been converted into suburban homes and cropland. But we simply do not know how many species have dwindled as the cloud forests of Central America have been replaced by coffee groves and cattle pastures, or as the lowland rainforests of Indonesia and Malaysia have become oil palm and pulpwood plantations.[18]

The single largest factor behind the declines of species is loss of habitat due to disturbance resulting from peoples' actions. In some regions, habitat loss has become so extensive that not just individual species but entire communities face extinction. The laurel forests of Colombia's inter-Andean valleys, the heathlands of western Australia, the unique seasonally dry forest of the Pacific island of New Caledonia—all have been overrun by humankind's activities. Some such habitats were naturally localized due to unique geologic or climatic conditions, but others once covered vast expanses. In the United States, 30 million hectares of forest along the southern Atlantic and Gulf coastal plain were previously dominated by fire-adapted longleaf pine communities. Logging, land clearance, and fire suppression have reduced these unique forests by at least 85–90 percent, pushing a number of animal and plant species that inhabit longleaf pine communities onto the U.S. endangered species list.[19]

Environmental transformation has been equally intensive in the southeast corner of the state of Florida, where several unique plant communities, such as subtropical hardwood hammocks and limestone ridge pinelands, now exist as a handful of isolated patches amid a sea of suburban homes, sugarcane fields, and citrus orchards. These irreplaceable remnants of what southeast Florida once was now require constant human vigilance to beat back a siege of nonnative plant interlopers, Brazilian pepper and Australian casuarina among them. Invasive species like these spread to new geographic regions, usually with intentional or inadvertent human help, and, finding their new location to their liking, then proceed to crowd out native flora and fauna. In the United States, invasive plants infest at least 44 million hectares and are expanding rapidly. In certain susceptible habitats, such as long-isolated oceanic islands, subtropical heathlands, and temperate grasslands, controlling invasives is the single biggest challenge for natural resource managers.[20]

South Africa faces one of the largest invasive species problems of any nation and has a great deal at stake: the fynbos heathlands and montane forest of the country's Cape region hold more native plant species—8,600, most of them endemic—in a smaller area than any other place on Earth. South Africans are increasingly aware of the threat that exotics pose, and in 1996 the government initiated a labor-intensive program to fight invasives with hacksaw and hoe. Some 40,000 people are employed to cut and clear Australian eucalypts, Central American pines, and other unwanted guests in natural areas. Even the government's water department has become a sponsor of invasive control, since many non-native trees are heavy drinkers, with deep root systems that tap into scarce groundwater supplies. It is a measure of the scale and severity of the invasive problem that this control effort is South Africa's single largest public works program.[21]

As if invasive species alone were not enough of a problem, other large-scale ecological alterations can multiply their negative effects in unpredictable and damaging ways. For instance, much of the world is now saturated in nitrogen compounds because of our overuse of nitrogen-based synthetic fertilizers and fossil fuels and our promotion of deforestation and land clearance. (Nitrogen is required by all plants for growth and development. In elemental form it is by far the most abundant gas in Earth's atmosphere, but it cannot be used by plants until "fixed" in organic form, a job performed in nature primarily by microorganisms.) Studies of North American prairies found that the plants that responded best to excess nitrogen tended to be weedy invasives, not the diverse native prairie flora. Likewise, native plant and animal species already pressed for survival in greatly reduced habitats may additionally have to contend with altered rainfall patterns, temperature ranges, seasonal timing, and other effects of global climate change.[22]

Already, scientists are detecting what could be the first fingerprints of an altered global atmosphere on plant communities. Data from tropical forest research plots worldwide indicate that the rate at which rainforest trees die and replace each other, called the turnover rate, has increased steadily since the 1950s. This suggests that the forests under study are becoming "younger," increasingly dominated by faster-growing, shorter-lived trees and woody vines—exactly the kinds of plants expected to thrive in a carbon dioxide-rich world with more extreme weather events. Without major reductions in global carbon emissions, forest turnover rates will likely rise further, and over time could push to extinction many slower-growing tropical tree species that cannot compete in a carbon-enriched environment.[23]

The ecological world emerging from these global trends is most striking in its greater uniformity. The richly textured mix of native plant communities that evolved over thousands of years now appears increasingly frayed, replaced in many regions by extensive areas under intensive cultivation or heavy grazing, lands devoted to settlements or industrial activities, and *secondary* habitats—partially disturbed areas dominated by shorter-lived species, and increasingly hosting high concentrations of invasive organisms. A 1994 mapping study by the organization Conservation International estimated that nearly three quarters of our planet's habitable land surface (that which is not bare rock, drifting sand, or ice) already is either partially or heavily disturbed.[24]

Habitat loss involves more than just the disappearance of wilderness expanses or pristine landscapes, however. Landscapes populated by traditional farming communities or nomadic pastoral groups commonly contain relatively diverse patchworks of small-scale cultivation, fallow fields, hedgerows, seasonal grazing areas, and managed forest patches of varying ages where people harvest wild plant products for household use, such as fuelwood and raw materials for craftwork. In addition, these areas shelter wildlife species and contain high concentrations of wild relatives of crop plants, many of which favor partially disturbed habitats. As people intensify their cultivation methods or grazing pressure in response to factors like local population growth or increased market demand for farm produce, they often transform diverse habitat mosaics into more uniform landscapes.[25]

The current mass extinction is even draining the gene pools that we ourselves are responsible for creating—the genetic diversity within cultivated plants. Traditional farmers developed a wealth of distinctive varieties within crops by selecting and replanting seeds, tubers, or cuttings from uniquely favorable plants—perhaps one that matured slightly sooner than others, was unusually resistant to pests, or possessed a distinctive color or taste. Over time, farmers selected thousands of folk varieties, or "landraces," within most major crops—until this century, when genetic erosion rather than diversification of crops has been the rule.[26]

Of Food and Farming

The value of biodiversity to humanity, and the potential ramifications of its loss, is nowhere more evident than in our food supply. Roughly one in every three plant species has edible buds, fruits, nuts, seeds, tubers, leaves, roots, or stems. During the nine tenths of human history when everyone lived by hunting and gathering, an average culture would probably have had knowledge of several hundred edible plant species. Even today, at least eight millennia after humans first began to practice agriculture, wild foods continue to supplement the diet of millions of rural residents worldwide. Tuareg women in Niger, for instance, regularly harvest desert panic-grass, shama millet, and other wild grains while migrating with their animal herds between wet- and dry-season pastures. In rural northeast Thailand, wild foods gathered from forests and field margins make up half of all food eaten by villagers during the rainy season. In the city of Iquitos in Amazonian Peru, fruits of nearly 60 species of wild trees, shrubs, and vines are sold in the city produce markets. Residents in the surrounding countryside are estimated to obtain a tenth of their entire diet from wild-harvested fruits.[27]

In most instances, the importance of wild plant and animal foods is measured in more than just calories. Part of their value has to do with timing—people often rely heavily on wild foods during seasonal periods of food scarcity in between agricultural harvests, or as foods eaten to stave off famine when a drought, flood, or other disaster strikes. In addition, wild foods are valuable sources of vitamins, trace minerals, and other essential nutrients. The Maasai, a cattle-herding pastoral people in Kenya and Tanzania, have been described as having the "world's worst diet"—they consume primarily milk products and meat, and obtain up to two thirds of their calories from animal fat. Nonetheless, most Maasai have low blood cholesterol levels and do not suffer from heart disease or other diet-related health problems, a fact researchers attribute to the wild plants the Maasai regularly add to their soups and stews, chew as gums and resins, or consume as medicinal tonics. A number of the wild-plant products most favored by the Maasai, such as acacia tree bark (added to soups), have been shown in biochemical trials to be quite effective at removing cholesterol from the body.[28]

The ancient shift from hunting and gathering to settled agriculture occurred independently in many different regions, as people gradually began to live closer together, became less nomadic, and focused their food and fiber production on plants that were amenable to repeated sowing and harvesting by human hands, or that could be tended for their tubers, fruits, or other products. In the 1920s, the legendary Russian plant explorer Nikolai Vavilov identified geographic centers of crop diversity, including Mesoamerica, the central Andes, the Mediterranean Basin, the Near East, highland Ethiopia, and eastern China. He inferred correctly that most of these "centers" correspond to where crops were first domesticated. (See map.) For instance, native Andean farmers not only brought seven different species of wild potatoes into cultivation, they also domesticated common beans, lima beans, passion fruit, quinoa and amaranth (both of which have grain-like seeds), and a host of tubers, roots, and other crops little known outside the Andes, such as *oca, ullucu,* and *tarwi*—more than 70 species of food plants in all. In addition, certain regions are centers of diversity for crops that were not originally domesticated there. This is the case with corn in the Andes (it was originally domesticated in Mexico), and with wheat and barley in highland Ethiopia. Neither grain is native to Sub-Saharan Africa, but Ethiopian farmers have grown them for thousands of years, assembling extremely diverse varietal mixes.[29]

Subsistence farmers have always been acutely attentive to selecting for the differences among varieties of the same crop, in part because those differences can help them cope with variability in their environment. Peasant farmers in Mexico who grow maize in rainfed fields commonly sow several different varieties that vary in the length of time and amount of water they need to reach maturity. In this way, farmers reduce the risk of catastrophic crop failure and increase their odds of obtaining a sufficient harvest whether the annual rains arrive early, late, sporadically, or in excess. In other instances, people may favor varietal diversity primarily for esthetic reasons. Native Hawaiians and other Pacific Islanders, for instance, may well have selected for distinctiveness in their crops based on concepts of ceremonial importance and beauty, much as Europeans have prized similar qualities in different varieties of roses they selected over the centuries.[30]

By simply paying close attention to differences among and within the crops they grow, farmers have selected and developed an impressively rich legacy of crop varietal diversity. Earlier this century, for instance, there were probably over 100,000 rice landraces under cultivation in Asia, with at least 30,000 in India alone. Among certain indigenous communities in Indonesia, each family maintains its own distinctive suite of rice varieties, creating complex patterns of varietal diversity even within individual villages. Rice is also one of the few crops to have been domesticated on more than one continent. Working with a completely different rice species, farmers in West Africa selected and diversified African rice into thousands of indigenous varieties.[31]

On-farm crop selection remains vital in developing countries, where farmers continue to save 80–90 percent of their own seed supplies (including both landraces and professionally bred seeds acquired initially through government extension programs or local markets). In industrial nations, however, the seed supply process has become increasingly formalized and centralized during this century. Professional plant breeders (working both in government stations and in private industry) have taken up the job of crop improvement, while the responsibility for developing, replicating, distributing, and saving seeds has been assumed by government agricultural programs and private companies. In recent decades, there has been a steady decline in public-sector funding of most crop improvement programs (a trend evident across

MAP

Original Regions of Domestication of Selected Crops

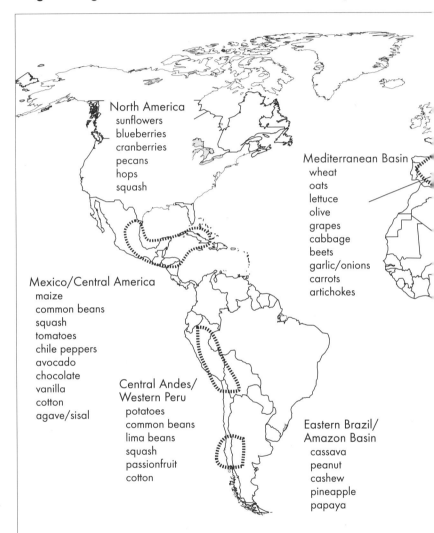

North America
sunflowers
blueberries
cranberries
pecans
hops
squash

Mediterranean Basin
wheat
oats
lettuce
olive
grapes
cabbage
beets
garlic/onions
carrots
artichokes

Mexico/Central America
maize
common beans
squash
tomatoes
chile peppers
avocado
chocolate
vanilla
cotton
agave/sisal

Central Andes/
Western Peru
potatoes
common beans
lima beans
squash
passionfruit
cotton

Eastern Brazil/
Amazon Basin
cassava
peanut
cashew
pineapple
papaya

Note: Vavilov centers of origin are regions where particularly high
concentrations of crops were first domesticated. Note that a number
of crops were first domesticated outside of Vavilov's centers, particularly in North America, eastern South America, and Sub-Saharan
Africa.

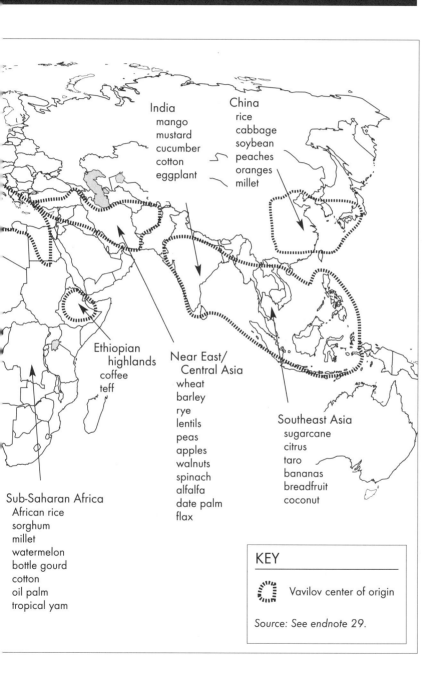

India
mango
mustard
cucumber
cotton
eggplant

China
rice
cabbage
soybean
peaches
oranges
millet

Ethiopian
highlands
coffee
teff

Near East/
Central Asia
wheat
barley
rye
lentils
peas
apples
walnuts
spinach
alfalfa
date palm
flax

Southeast Asia
sugarcane
citrus
taro
bananas
breadfruit
coconut

Sub-Saharan Africa
African rice
sorghum
millet
watermelon
bottle gourd
cotton
oil palm
tropical yam

KEY

Vavilov center of origin

Source: See endnote 29.

nearly all areas of agricultural research, save biotechnology), while private-sector crop research programs have received increased investment. Seeds are in effect being transformed from a public good to a private commodity.[32]

The power and promise of scientifically based plant breeding was confirmed by the 1930s, when the first commercial hybrid corn was marketed by the Hi-Bred Company (now known as Pioneer Hi-Bred) under its founder (later the U.S. Secretary of Agriculture), Henry A. Wallace. Hybrid varieties are easier to develop in cross-pollinated plants like corn or squashes, but are less readily achieved in self-pollinated crops like wheat. Hybrids that are widely grown by farmers tend to be especially high yielding (the bottom line for commercial farming) due to a biological phenomenon known as *heterosis*, or hybrid vigor. Seed supply companies prefer hybrids because they are more profitable, selling for three to eight times more than standard, "open-pollinated" varieties. Moreover, second-generation seeds from hybrids do not retain the traits of their parents, ensuring that farmers must purchase hybrid seed anew from the supplier rather than saving their own stock.[33]

Over the past two decades, some farmers have also been disenfranchised from seed saving. Under certain European Union and United States laws, it is illegal for farmers to save and replant seed from plant varieties that have been registered by breeders or other seed suppliers. In addition, genetic engineers have been hard at work identifying methods to short-circuit seed saving biologically that are more flexible and more widely applicable than hybridization. The most prominent and controversial development to date has been a genetically engineered "terminator" gene sequence which, once inserted into the DNA of a plant, allows scientists to control the viability of its seeds by using a chemical trigger. In effect, exposing a plant to the selected chemical sets off a genetic chain reaction that causes the plant to produce a toxic substance in its seeds, killing the embryo inside. The terminator technology has been patented jointly by the U.S. Department of Agriculture and a U.S.-based cotton seed supplier, Delta and Pine Land Company (which was recently swallowed up by the agribusiness giant, Monsanto). If the technology is developed commercially, it would enable seed suppliers to control the reproduction of self-pollinated crops currently not amenable to hybrid production, such as wheat and rice. The terminator gene complex is by no means an isolated example; nearly all major seed companies have a similar kind of genetic viability control process in the research or development stage.[34]

New agricultural biotechnologies are being put to use to control more than just seed viability, of course; with genetic engineering, scientists are able to manipulate species in ways that resemble medieval alchemy. Plant breeders can now move genes between completely unrelated organisms, such as a gene for cold tolerance that was transferred from a flounder (a fish that dwells on the sea floor) to a tomato. The adoption of transgenic crops has expanded spectacularly over the past five years, particularly in soybeans, maize, tobacco, cotton, and canola (an oil crop also known as rapeseed). The most common genetically engineered trait to date has been herbicide tolerance, followed by insect and virus resistance. Despite its obvious power, however, genetic engineering is still a long way from making conventional plant breeding obsolete. The most desirable plant traits such as yield potential are controlled by intricate complexes of genes that can be transferred between crop lines only through conventional breeding techniques. Genetic engineering methods, by contrast, transfer relatively simple traits controlled by one or a few genes.[35]

Beneath the veneer of modern agro-industry, the productivity of our global food supply still depends on the plant diversity in wildlands and traditional agricultural landscapes. Overall, about half of the astounding yield gains realized in mechanized agriculture this century have come from improvements that plant breeders have made to crop varieties. (The other half of the gains is due to changes in agronomic practices, primarily increased inputs of nitrogen fertilizer.) Much of the genetic architecture of high-yielding crop varieties assembled and shaped by professional plant breeders can be traced back to traditional landraces. Moreover, the flow of useful genes into breeding lines is ongoing. The typical commercial "lifespan" of a professionally bred crop variety is only five to 10 years, after which time pest and disease problems, in particular, become increasingly difficult to manage, leading farmers and breeders to seek new, more resistant, or more productive varieties. According to one study reported by economist Timothy Swanson, each year, plant breeders return to landraces and their wild relatives for about 6 percent of the germplasm lines used in breeding new crop varieties (the remainder are professionally bred varieties and advanced breeding lines previously developed by breeders). For major grain crops like wheat and rice, the estimated annual infusion of landrace and wild relative germplasm is about 15 percent. Although small, these percentages reflect breeders' continuing reliance on landraces and wild crop relatives for specific traits like disease resistance and palatability that are simply not present in the easier-to-work-with advanced breeding lines.[36]

The value of these traits to global agriculture is measured in the billions of dollars. Modern commercial food production, where crops are grown on short rotations in large expanses, and where pests can migrate to new regions more easily than ever before, places unique stresses on crop plants, which breeders have overcome by turning in many cases to the gene pools of wild relatives. Imagine giving up sugarcane, strawberries, tomatoes, and wine grapes; none of these crops could be grown commercially without the genetic contributions of their respective wild relatives. During the mid-19th century, a non-native aphid-like insect and nonnative fungus began devastating vineyards throughout Europe, but three wild North American grape species proved resistant to the pests. All European wine grapes are now grown exclusively on North American rootstocks.[37]

Traditional crop varieties are equally indispensable for global food security. Subsistence farmers around the world continue to grow primarily either landraces or locally adapted versions of professionally bred seed. Such small-scale agriculture produces 15–20 percent of the world's food supply, is predominantly performed by women, and provides the daily sustenance of roughly 1.4 billion indigenous and peasant farmers. Moreover, as previously mentioned, genetic traits borrowed from landraces are key building blocks of the professionally bred crop varieties that feed the rest of us. For instance, a popular wheat variety called "Marquis" at one point accounted for about 90 percent of the North American Great Plains spring wheat acreage; it originated from a cross between an Indian wheat landrace named "Hard Red Calcutta" and a traditional European variety called "Red Fife." A landrace from Turkey is a parent of many of the wheat varieties now grown in the northwestern United States, to which it has provided crucial genetic resistance to rusts, smuts, and other fungal diseases afflicting wheat.[38]

As we enter the next millennium with an expanding human population placing increasingly intensive demands on natural resources, agricultural biodiversity faces an uncertain future. Conversion of wild habitats can reduce the availability of wild foods, and threatens populations of wild relatives of crops. In the United States, for instance, a recent study by the Center for Plant Conservation revealed that about two thirds of all rare and endangered plants are close relatives of cultivated species. If these species go extinct, a pool of potential genetic benefits for global agriculture will also vanish.[39]

For some plants, it is changes within agricultural landscapes that give cause for concern. The closest wild ancestor of corn is one such species, a lanky, sprawling annual grass called teosinte. Native to Mexico and Guatemala, teosinte occurs in eight geographically separated populations, each of which bears unique genetic characteristics. Botanist Garrison Wilkes of the University of Massachusetts regards seven of these populations as rare, vulnerable, or already endangered—primarily due to the decline of traditional agricultural practices, and to increased livestock grazing in the field margins and fallow areas favored by teosinte, with regional

increases in human numbers being an underlying cause of both problems. Overall, teosinte is not yet threatened with extinction. But because the plant is so closely related to domesticated corn, every loss of a unique teosinte population reduces a source of genes for breeding better-adapted corn plants.[40]

There is also grave concern for the old crop landraces. By volume, the world's farmers now grow more sorghum, string beans, sweet potatoes, and other crops than ever before, but they grow fewer varieties of each crop. Particularly in industrial nations, crop varietal diversity has undergone a massive turnover this century; the proportion of varieties grown in the United States before 1904 but no longer present in either commercial agriculture or a major seed storage facility ranges from 81 percent for tomatoes to over 90 percent for peas and cabbages. Figures are less comprehensive for developing countries, but China is estimated to have gone from growing 10,000 wheat varieties in 1949 to only 1,000 by the 1970s, while only 20 percent of the corn varieties cultivated in Mexico in the 1930s can still be found there—an alarming decline for the cradle of corn domestication.[41]

Crop varieties are lost for many reasons. Sometimes an extended drought or civil conflict destroys harvests and farmers must consume their stocks of planting seed just to survive. Long-term climate change has also been implicated as a trigger for varietal abandonment. In Senegal and Gambia, two decades of below-normal rainfall created a growing season too short for traditional rice varieties to produce good yields. When fast-maturing rice cultivars became available through development aid programs, local farmers rapidly switched to them because of the greater short-term harvest security the new cultivars offered.[42]

In the majority of cases, however, farmers have abandoned traditional seeds as they have adopted new varieties or changed agricultural practices. Some have moved out of farming altogether. In industrial countries, crop diversity has declined in concert with the steady commercialization, consolidation, and depopulation of agriculture this century:

fewer family farmers, and fewer seed companies offering fewer varieties for sale, eventually will translate into fewer crop varieties planted in fields or saved after harvest. The commercial seed supply business is now dominated by multinational corporations that not only sell seeds but also manufacture pesticides, fertilizers, and other agricultural inputs. Not surprisingly, some companies now breed seeds specifically adapted to use the other products they sell. Monsanto's best-selling line of genetically engineered seeds goes by the brand name of "Roundup Ready®," because they contain a gene for resistance to Roundup, a popular herbicide also produced by Monsanto.[43]

Only 20 percent of the corn varieties cultivated in Mexico in the 1930s can still be found there.

In most developing countries, varietal losses accelerated in the 1960s, when the famed international agricultural development program known as the Green Revolution introduced new strains of wheat, rice, and other major crops bred to produce much higher yields than traditional varieties. Developed to boost food self-sufficiency in famine-prone countries, the high-yielding Green Revolution varieties were widely distributed, often with government subsidies to encourage their adoption, and they displaced landraces from many farmland areas.[44]

In many places—most notably the world's intensively cultivated "breadbasket" and "ricebowl" regions of North America, Europe, and Asia—crops now exhibit what the U.N. Food and Agriculture Organization (FAO) politely calls an "impressively uniform" genetic base. A survey of nine major crops in the Netherlands found that the three most popular varieties for each crop covered 81–99 percent of the respective areas planted, a pattern evident in most other industrialized nations as well. Such patterns also emerged on much of the developing world's prime farmland by the 1970s. One single wheat variety blanketed 67 percent of Bangladesh's wheat acreage in 1983 and 30 percent of India's the following year.[45]

The ecological risks we take in adopting genetic uniformity are enormous, and keeping these risks at bay requires an extensive infrastructure of agricultural scientists, extension workers, and credit programs—as well as frequent applications of pesticides and other potent agrochemicals. A particularly heavy responsibility falls on professional plant breeders, who are now engaged in a relay race to develop ever more robust crop varieties before those already being grown commercially succumb to evolving pests and diseases or changing environmental conditions.[46]

Breeders started this race earlier this century with a tremendous genetic endowment at their disposal, courtesy of nature and generations of subsistence farmers. Despite major losses, this wellspring is still far from running dry. Estimates are that plant breeders have used only a small fraction of the varietal diversity present in crop gene banks (facilities that store seeds under cold, dry conditions which can maintain seed viability for decades).[47]

At the same time, even with genetic engineering we can never be sure that what is already stored will cover all our future needs. When grassy stunt virus began attacking Asian rice fields in the 1970s, breeders located genetic resistance to the disease in but a single collection of one population of the wild rice *Oryza nivara* in Uttar Pradesh, India—and that population has never been found again since. Such experiences make it clear that achieving global food security will be illusory unless we can reinvigorate biodiversity in agricultural landscapes.[48]

Of Medicines and Everyday Goods

In a doctor's office in Germany, a man diagnosed with hypertension is prescribed reserpine, a drug derived from the Asian snakeroot plant. In a small town in India, a woman complaining of stomach pains visits an Ayurvedic healer, and receives a soothing and effective herbal tea as part of her treatment. In a California suburb, a headache sufferer unseals a bottle of aspirin, a compound originally isolated from European willow trees.[49]

People everywhere rely on plants for staying healthy and extending the quality and length of their lives. One quarter of the prescription drugs marketed in North America and Europe contain active ingredients derived from plants. Plant-based drugs are part of standard medical procedures for treating heart conditions, childhood leukemia, lymphatic cancer, glaucoma, and many other serious illnesses. (See Table 3.) Some plant-based drugs have been prescribed by physicians for more than a century—digitalis is one, derived from the European foxglove plant and used to treat congestive heart failure. Worldwide, the over-the-counter value of plant-derived drugs is estimated at more than $40 billion annually. Major pharmaceutical companies and research institutions such as the U.S. National Cancer Institute invest in extensive plant-screening programs as a means of identifying new drugs.[50]

The World Health Organization (WHO) estimates that 3.5 billion people in developing countries rely on plant-based medicine for their primary health care. Ayurvedic and other traditional healers in South Asia use at least 1,800 different plant species in treatments and are regularly consulted by some 800 million people. In China, where medicinal plant use goes back at least four millennia, healers have employed more than 5,000 plant species. At least 89 plant-derived commercial drugs used in industrial countries were originally discovered by folk healers, many of them women. Traditional medicine is particularly important for rural residents who typically are not well served by formal health care systems. Recent evidence suggests that when economic woes and structural adjustment programs restrict governments' abilities to provide health care, urbanized middle-class residents of developing countries also turn to more affordable traditional medical experts.[51]

Traditional herbal therapies are growing rapidly in popularity in industrial countries as well. The FAO estimates that

TABLE 3

Examples of Drugs Derived from Plant Compounds

Drug	Medical Uses	Source Plant and Origin
Codeine	Used for pain relief and as cough suppressant.	poppy flower (Asia)
Pilocarpine	Treatment for eye disease (glaucoma).	*jaborandí* plant (Amazon Basin)
Tubocurarine	Used to relax skeletal muscles during surgical procedures.	*curare* vine (South America)
Scopolamine	Prevents motion sickness, also used as sedative.	*Datura* plant (Americas)
Gamma-linolenic Acid	Used to treat eczema, diabetic nerve damage, and pancreatic cancer.	evening primrose (North America)
Colchicine	Anti-tumor agent used in cancer treatment, also used to treat gout.	autumn crocus (Eurasia)
Quinine, Quinidine	Anti-malarial agent, also used to treat heart arrhythmia.	*Cinchona* tree (South America)
Camphor	Relieves rheumatic pain.	camphor tree (Asia)
Pseudo-ephedrine	Bronchodilator, also used to treat rhinitis.	*ma-huang* shrub (China)
Taxol	Anti-tumor agent used in cancer treatment.	Pacific yew tree (North America)

Source: See endnote 50.

between 4,000 and 6,000 species of medicinal plants are traded internationally, with China accounting for about 30 percent of all such exports. In 1992, the booming U.S. retail market for herbal medicines reached nearly $1.5 billion, and the European market is even larger.[52]

Despite the demonstrable value of forests and other

habitats, their alteration by humans frequently eliminates sites rich in wild medicinal plants. This creates an immediate problem for folk healers when they can no longer find the plants they need for performing certain cures—a problem commonly lamented by indigenous herbalists in eastern Panama, among others. Moreover, strong consumer demand and inadequate oversight of harvesting levels and practices mean that wild-gathered medicinal plants are commonly overexploited.[53]

The bark of the African cherry tree, for example, is highly esteemed by traditional healers in many central and West African countries, but European doctors also now value it as a principal treatment for prostate disorders. In recent years the country of Cameroon has been the leading supplier of African cherry bark to European countries and other international markets, whose imports averaged over 3,000 tons annually in the early 1990s. Unfortunately, clearance of the tree's montane forest habitat in Cameroon, combined with the inability of the government forestry administration to manage the harvest, has led to widespread, wanton destruction of cherry stands.[54]

A similar problem affecting both national and international consumers has emerged in Mexico with a herb called valerian (*valeriana* in Spanish) that is valued for the calming effect of its pungent roots. Domestic demand for valerian was already strong enough to place heavy harvesting pressures on wild populations, and since the early 1990s additional demand for the plant has arisen in Europe and the United States. Depletion of valerian has become so severe that medicinal plant traders in Mexico now often substitute a related but inferior species because the most powerful valerian species is not available.[55]

In addition to the immediate losses, every dismantling of a unique habitat represents a loss of future drugs and medical compounds, particularly in species-rich habitats like tropical forests. To date, fewer than 1 percent of all plant species have been screened by chemists to see what bioactive compounds they may contain. And the nearly 50 drugs already derived from tropical rainforest plants are estimated to represent only about 12 percent of the medically useful compounds waiting to be discovered in rain forests.[56]

Alarmingly, many rural societies are rapidly losing their cultural knowledge about medicinal plants. In communities undergoing accelerated westernization, fewer young people are interested in devoting themselves to the extensive training required to learn about traditional healing plants and how to use them. From Samoa to Suriname, most herbalists and healers are elderly, and few have apprentices studying to take their place. Ironically, as this decline has accelerated, there has been a resurgent interest in ethnobotany—the formal study of how people classify, conceptualize, and use plants—and other fields of study related to traditional medicinal plant use. Ethnobotanists surveying medicinal plants used by different cultures are racing against time to document traditional knowledge before it vanishes along with its last elderly practitioners.[57]

For the one quarter of humanity who live at or near subsistence levels, plant diversity offers more than just food security and medicines—it also provides material for the roof over their heads, fuel to cook their food, utensils to eat with, in sum about 90 percent, on average, of their material needs. Consider the utility of palms: temperate zone dwellers may think of palm trees primarily as providing an occasional coconut to eat or the backdrop to an idyllic island vacation, but tropical peoples have a different perspective. (See Table 4.) The babassu palm from the eastern Amazon Basin, for instance, has more than 35 different uses—construction material, fiber source, medicine, game attractant, edible oil source, even insect repellent. Commercial extraction of babassu products is a part- or full-time economic activity for more than 2 million rural Brazilians in the north and northeast regions of the country.[58]

Indigenous peoples throughout tropical America have been referred to as "palm cultures." The posts, floors, walls, and beams of their houses are made from the wood of palm trunks, while the roofs are thatched with palm leaves. They

use baskets and sacks woven from palm leaves to store household items, including food—which may itself be palm fruits,
palm hearts (the young growing shoot of the plant), or wild
game hunted in palm stands (where animals congregate to
feed on palm fruits) and killed with spears and arrows made
from palm stems and leaves. At night, family members usually drift off to sleep in hammocks woven from palm fibers.
When people die, they may be buried in a coffin carved from
a palm trunk.[59]

Palms are exceptionally versatile, but they are only part
of the spectrum of useful plants in highly
diverse environments. In northwest
Ecuador, indigenous cultures that practice shifting agriculture use over 900
plant species to meet their material, medicinal, and food needs. Halfway around
the world, Dusun and Iban communities
in the rain forests of central Borneo
employ a similar total of plants in their
daily lives. People who are more integrated into regional and national economies
tend to use fewer plants directly, but still
commonly depend on plant diversity for
household uses and to generate cash
income. In India, at least 5.7 million peo-
ple make a living harvesting non-timber forest products, a
trade that accounts for nearly half the revenues earned by
Indian state forests. All told, the number of materially useful
plant species that enter into regional trade networks in
developing countries probably totals in the tens of thousands.[60]

Fewer than 1 percent of all plant species have been screened by chemists to see what bioactive compounds they contain.

Those of us who live in manicured suburbs or urban
concrete jungles may meet more of our material needs with
metals and plastics, but plant diversity still enriches our lives.
Artisans who craft musical instruments or furniture, for
instance, value the unique acoustic qualities and appearance
of the different tropical and temperate hardwoods that they
work with—aspects of biodiversity that ultimately benefit

TABLE 4

Examples of Useful Palms

Name	Region Found	Common Uses
Sago Palm	Indonesia, Malaysia, Philippines, New Guinea	Stem stores large amounts of edible starch, which is extracted as a staple food by forest peoples, and also sold in regional markets. Leaves used as thatching, stem wood for construction purposes.
Sugar Palm	India, Bangladesh, Sri Lanka	Flower bud is tapped for a sap that is dried into sugar. Leaves used for weaving baskets and fishnets. Stem produces edible starch, and palm heart also eaten.
Rattan Palms (500+ species)[1]	Southeast Asia, Africa	Stems harvested for fashioning into cane furniture, a global industry. Some rattans also harvested commercially for palm heart. Rattan stems also used locally for fashioning baskets, mats, fish traps, and many other products.
Borassus Palms (2 species)[1]	Sub-Saharan Africa and India (arid regions)	All parts of palm used, especially leaves for thatching and basket weaving, and stem for house construction. Stem is tapped to make palm wine.
Assaí Palms (3 species)[1]	Northern and eastern South America	Produces a highly sought-after fruit, and is the leading palm harvested by the palm heart industry in Latin America.
Tagua Palms (3 species)[1]	Panama to Peru	Hard white seeds are called "vegetable ivory" and can be carved into figurines and tourist curios. Historically supported a large button-making industry, centered in Ecuador. Leaves used locally for thatch.

Table 4 *(continued)*

Mauritia Palm	Amazon Basin and northern South America	Fruits eaten fresh and dried, sold in regional markets; string from young leaves used to make rope, hammocks, baskets, and other household items; spongy core of leaf blade can be used to make mats and for paper production.
Xate Palm	Mexico, Guatemala, Belize	Leaves harvested and exported for the international floral trade.
Carnauba Palm	Brazil	Produces a premium vegetable wax, traded internationally.

[1]Species totals refer only to species of palms commonly used, not related unharvested species.

Source: See endnote 59.

anyone who listens to classical music or purchases handcrafted furniture. Among the non-food plants traded internationally on commercial levels are at least 200 species of timber trees, 42 plants producing essential oils, 66 species yielding latexes or gums, and 13 species used as dyes and colorants.[61]

Another category of plants that enter into wide international trade are those valued for their ornamental or horticultural qualities. The most heavily traded ornamental plants are cacti, with annual trade recorded in the Convention on International Trade in Endangered Species (CITES) statistics averaging around 13 million individual cactus plants. (Unrecorded trade is thought to be substantially higher.) Other widely traded kinds of horticultural plants include lilies, which come primarily from Turkey and the Middle East; orchids, primarily from Latin America and Southeast Asia; and desert-adapted succulent plants from southern Africa and Madagascar that bear a superficial resemblance to cacti. The vast majority of ornamental plants are propagated in captivity, but there continues to be a substantial wild trade, particularly of species that are hard to obtain or slow growing. Mexico, for instance, exports around 50,000 cacti annually, of which virtually all are thought to be wild-harvested, despite domestic laws prohibiting such collection. Cactus aficionados in Europe and Japan are so eager to add to their holdings that it is not uncommon for news of the discovery of a new cactus species in Mexico to trigger intensive collecting of the species from the location where it was discovered. In the case of several rare endemic species, such collecting has led to their near extinction shortly after their discovery by science.[62]

As with medicinals, the value that plant resources have for craft production, industrial use, or household needs has often not prevented their local or regional decline. One of the most valuable non-timber forest products is rattan, a flexible cane obtained from a number of species of tropical vine-like palms that can grow up to 185 meters long as they scramble from the shadowy forest floor up to the sunlit forest canopy. Asian rattan palms support an international furniture-making industry worth $3.5–$6.5 billion annually. Unfortunately, many Asian rattan stocks are declining because of forest clearance and overharvesting. In the past few years, some Asian furniture makers have even begun importing rattan supplies from Nigeria and other central African countries.[63]

On a global level, declines of wild plants related to industrial crops such as cotton or plantation-grown timber could one day limit our ability to cultivate those commodities by shrinking the gene pools needed for breeding new varieties. Many other highly sought wild plants are not easily grown as crops, and have become increasingly scarce when governments have been unable to manage the wild harvest. American mahogany, for instance, is arguably the most highly prized tropical hardwood, but it cannot be grown in large plantations because of devastating attacks by shoot borers, an insect that damages the growing tip of the young tree. The international woodworking industry continues to rely on wild-harvested trees, which once grew across a vast area of the American tropics, from Mexico and Florida to Brazil and Bolivia. Woodworkers now pay more money than ever before for mahogany of declining quality—the end result of an international industry unable to change course after decades of unsustainable harvest and consumption.[64]

More locally, declines of materially useful species mean life gets harder in the short term. When a tree species favored for firewood is overharvested, women must walk farther to collect their family's fuel supply, make do with an inferior species that does not burn as well, spend scarce money purchasing fuel from vendors, or devote limited land and labor to growing a fuelwood crop. When a fiber plant collected for sale to handicraft producers becomes scarce, it is harder for collectors to earn an income that could help pay school fees for their children. Whether we are rich or poor, plant diversity enhances the quality of our lives—and many people already feel its loss.

Ecological Services

Most assessments of the value of plant diversity (like this paper so far) tend to focus on the benefits provided to human welfare by individual species and varieties. These benefits, on the whole, are relatively easy to quantify—for instance, the economic boost provided to the U.S. tomato industry by new genes introduced from a wild Peruvian tomato species in the 1970s has been estimated at $8 million per year.[65]

The benefits provided by broader elements of plant diversity, such as intact natural communities and well-running ecosystems, are more diffuse and harder to apply a price tag to, but no less crucial. To start with, communities and ecosystems provide habitat for the thousands of wild species that people find materially useful. But they also provide a wide range of "natural services" in their own right. Many natural services work across entire landscapes, as when a forest in the headwaters of a river valley absorbs a heavy rainfall and releases the runoff gradually to an irrigation project downstream, or when annual flooding in the lower reaches of a free-flowing river brings in nutrient-rich sediment, rejuvenating the fertility of fields for planting once the waters subside. Migratory animals, too, provide benefits across great distances. Long-nosed bats move 3,000 miles seasonally up and down the deserts of Mexico and the southwestern United States, pollinating century plants that are harvested for the tequila industry, and columnar cacti that provide choice edible fruits.[66]

Other ecological services unfold on a smaller, more intricate scale, like the benefits gained from the way farmers compose fields, orchards, and garden plots in subsistence farming systems. Often a number of different crops are combined or intercropped together in a single field. Estimates are that nearly 80 percent of farmland in West Africa is intercropped, as is the majority of bean production in Latin America. This strategy favoring combinations of crops, or "polycultures," usually reaches its peak in the dooryard gardens immediately adjacent to farmers' homes. These accessible plots commonly contain upwards of 100 different species in half a hectare or less—a botanical riot of fruit trees, vegetable crops, herbs, gourds, spices, medicinals, and ornamental plants.[67]

Farmers gain a number of benefits from intercropping. Maintaining a row of sturdy fruit trees along the margins of a field can provide a windbreak for more delicate grain crops, preventing wind damage. Planting a low, fast-maturing crop together with a taller, longer-maturing crop can provide weed control as well as optimal use of soil nutrients—the fast-maturing crop will peak in its nutrient demands early, allowing the slower-maturing crop to flourish later on. A variety of different crop plants grown together may also support greater populations of beneficial insects that prey on crop pests, thereby reducing the need for chemical controls. Of course, a more diverse polyculture may also offer more ecological niches in which diseases can thrive, but the likelihood of any one disease breaking out in epidemic levels is reduced, since no disease is likely to affect all crops equally. In the end, subsistence farmers probably value most the stability and resilience of polycultures. As agricultural geographer Paul Richards puts it, "where a farmer in Europe borrows from the bank to tide him over a bad year...the West African farmer may not survive to try again."[68]

When subsistence farmers do plant an entire field to a single crop, they will usually switch crops for the next growing season or perhaps even leave the land unplanted, or "fallow," for several years—a strategy that utilizes the services of diversity over time rather than in space. Farmers in the high Andes mountains of Venezuela, for instance, have traditionally grown their staple crops (primarily potatoes, wheat, and beans) on a given plot of cleared land for only one to three years, then let the plot lie fallow for an extended period—usually eight to 20 years—before cultivating it again. (See Figure 2.)[69]

This spatial and temporal rotation of farm plots is a variation of shifting (also called swidden or "slash and burn") cultivation. (The burning step prior to cultivation, which serves to make nutrients in woody vegetation available to the newly planted crops, is not generally practiced in the high Andes, where the predominant vegetation consists of grasses, herbs, and small shrubs.) Shifting agriculture is practiced in some form or other on nearly 3 billion hectares of land in developing countries by an estimated 1 billion people. It not only works well in the high Andes in its traditional form, it is also the most effective and ecologically sound method yet devised for raising crops on the nutrient-poor soils of lowland tropical forest.[70]

The practice of letting land lie fallow offers a number of ecological and agronomic benefits. The native vegetation that reclaims the cultivated plot adds organic matter and steadily restores soil fertility. In addition, populations of crop pests—nematodes, mildews, insects—that built up in the plot under cultivation usually diminish rapidly. In the Venezuelan high Andes, fallow periods under traditional practices were commonly long enough to permit the reestab-

FIGURE 1

Diagram of Traditional Crop-Fallow Rotation in Venezuelan Andes

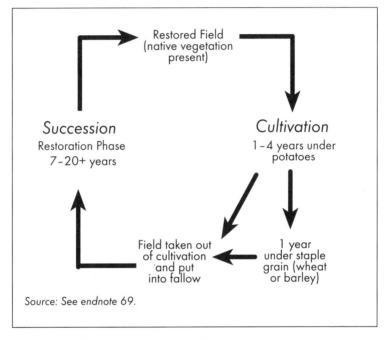

Source: See endnote 69.

lishment of the original vegetation cover, a unique community of low shrubs and rosette plants called *páramo*.[71]

The ecological benefits that community and landscape diversity provides become increasingly difficult to maintain the more we transform natural communities. Take the problem of habitat fragmentation, when a large, contiguous forest, savanna, or wetland area is reduced by human development to patchwork, island-like remnants of its former self. Natural islands in oceans or large lakes tend to be impoverished in species; their smaller area means they usually do not develop the ecological complexity that is characteristic of more extensive mainland areas. Moreover, when an island population of a species declines and disappears, adjacent mainland populations are likely to have trouble re-colonizing and replacing it, depending on how distant the island is.[72]

As a result, when large-scale agriculture, settlements, and roads sprawl across landscapes, remaining habitat fragments usually behave like the islands they have become: they lose species. In western Ecuador, the Río Palenque Science Center protects a square-kilometer remnant of the lowland rainforest that covered the region a mere three decades ago; now the center is an island amid cattle pasture and oil palm plantations. Twenty-four species of orchids, bromeliads, and other plants at Río Palenque have already succumbed to the "island effect" and can no longer be found there. One vanished species, an understory shrub, has never been recorded anywhere else and is presumed extinct.[73]

Even with these drawbacks, small areas of native vegetation can have enormous conservation value when they are all that is left of a unique plant community or rare habitat. But waiting to protect a community or habitat until only fragmented patches remain entails a high cost: smaller holdings require more intensive management than larger ones. In smaller reserves, managers often must simulate natural disturbances (such as prescribed burns to maintain fire-adapted vegetation); provide pollination, seed dispersal, and pest control services in place of vanished animals; reintroduce native species when they disappear from a site (perhaps due to a series of poor breeding seasons); keep out aggressive non-native species; and perform other duties the original ecosystem once did free of charge. Governments and societies that are unwilling or unable to shoulder these management costs will find that the biodiversity they intended to protect with nature reserves has vanished from within them.[74]

The diminishment of ecological services is felt in more than just wildland areas—it profoundly affects how we manage our food- and fiber-producing landscapes as well. In the latter half of this century, fallow periods in the Venezuelan Andes, like those of many agricultural regions worldwide, have been contracting and sometimes even vanishing altogether. During the 1960s, Venezuelan peasant farmers began to shift from growing primarily traditional subsistence crops to growing commercial varieties of potatoes, carrots, and gar-lic for sale to rapidly expanding urban markets. With strong market demand as an incentive, farmers boosted their pro-duction by adopting more intensive cultivation techniques, reducing fallow periods, and keeping larger portions of their lands under a limited number of crops for longer periods. Under these practices, not surprisingly, patchworks of fallow lands and native vegetation within agricultural areas of the high Venezuelan Andes have become less diverse over the past several decades.[75]

Of course, to replace the fertility restoration and pest control services that fallowing previously performed, farmers now apply fertilizers, pesticides, herbicides, and fungicides. Agricultural development programs sponsored by the Venezuelan federal government have given farmers ready access to chemical inputs, often at subsidized prices. With intensified production, subsidized costs, and strong market demand (Venezuela has a highly urbanized population), many farmers in the Venezuelan Andes have been able to boost their average household incomes substantially.[76]

These economic gains have exacted a price, however, for Venezuela's farmers, like farmers worldwide who have forsak-en ecological services, have been gambling with their health— and possibly that of consumers too. Many synthetic pesticides are suspected carcinogens and endocrine disruptors, and nei-ther the companies that manufacture them nor the govern-ments responsible for regulating them know their long-term effects on human health. Medical researchers in Sweden, for instance, recently found that exposure to several agrochemi-cals—including the world's most widely used herbicide, glyphosate, the active ingredient in Roundup—nearly tripled a person's odds of developing non-Hodgkin's lymphoma, one of the kinds of cancer increasing most rapidly in occurrence in the United States and other industrialized nations.[77]

Annual worldwide pesticide applications have risen steadily since mid-century, and these chemicals are now ubiquitous components of our environment. In Switzerland, rainfall regularly contains pesticides like atrazine and alochlor in such high concentrations that it often exceeds government safe drinking standards even before it hits the ground. While industrialized countries apply the biggest amounts of agrochemicals, residents of developing countries may face a greater danger of acute pesticide exposure. Farmworkers in developing countries commonly receive little training in safe pesticide application methods or appropriate dose levels, and many governments routinely permit the use of pesticides banned in industrial countries.[78]

Ecologists and agronomists have known for a long time that reducing the diversity of agricultural systems increases the magnitude of pest infestations, disease outbreaks, and other environmental problems. Scientists and other agricultural experts have done a remarkable job of boosting and maintaining agricultural productivity under these conditions, but not without creating long-term risks to the health both of ecological systems and of human beings. This and other unintended outcomes suggest that many of the end effects of heightened bio-uniformity are likely to appear unpredictably, or to manifest themselves in indirect ways through complex chains of events, making their impacts all the harder to plan for and mitigate.

Reducing the diversity of agricultural systems increases pest infestations, disease outbreaks, and other environmental problems.

Many farmers are deeply concerned about water pollution and the health hazards of the agrochemicals they use and are looking for alternatives, but there are no easy solutions. Governments concerned about the health of their citizens likewise face difficult choices. Expanding and intensifying agricultural production has enabled countries to feed and clothe a world population that has added 3 billion people in less than 50 years. Yet with human numbers now projected to climb from 6 billion to between 7.7 billion and 11.2 billion by 2050, we must somehow do it all over again—

while at the same time dealing with the costs of having forsaken the services of biodiversity the first time around. That challenge begins with saving the plant diversity that remains.[79]

Set Aside for Safekeeping

Broad recognition of the need to conserve plant resources is largely a twentieth century phenomenon. The first warnings about the global erosion of plant diversity were voiced in the 1930s by crop scientists such as Harry Harlan of the United States and Nikolai Vavilov, who realized the threat posed by farmers' abandonment of landraces in favor of new varieties that were spreading widely in an increasingly interconnected world. In the decades since their prophetic warnings (which went largely unheeded at the time), we have come to realize the importance of protecting plant biodiversity both off site in specialized institutions, such as botanical gardens and gene banks, and in native habitats and agricultural settings.[80]

The world's 1,600 botanical gardens hold extensive collections of non-domesticated species and ornamental plants. In all, botanical gardens tend representatives of tens of thousands of plant species—nearly 25 percent of the world's flowering plants and ferns, by one estimate. A significant percentage of species in botanical gardens, however, are represented by only a handful of individual plants.[81]

Most botanical gardens active today were established by European colonial powers to introduce economically important and ornamental plants throughout the far-flung reaches of empires, and to promote the study of potentially useful plants. Nowadays many botanical gardens are reorienting their mission toward species preservation, particularly in their research and education programs. Since the late 1980s, over 500 botanical gardens worldwide have coordinated efforts through a network called Botanic Gardens

Conservation International, which helps gardens establish and expand off-site conservation programs for rare plants. In the United States, the Center for Plant Conservation (CPC) has coordinated efforts among 28 botanical gardens to propagate endangered U.S. plants and reintroduce them into the wild. The CPC collections now total nearly 600 endangered U.S. plant species.[82]

Gene banks have focused primarily on storing seeds of the world's most important crop varieties and their immediate wild relatives. (One principal exception is the Royal Botanic Garden's Millennium Seed Bank in England, which holds more than 4,000 wild species and is expanding ambitiously toward a collection of one quarter of the world's flora.) Gene banks arose from the need of plant breeders to have readily accessible stocks of breeding material. Their conservation role came to the forefront in the 1970s, following mounting alarm among scientists over the loss of crop diversity in developing countries, coupled with unprecedented crop losses linked to genetic uniformity in the southern U.S. corn crop in 1970 and the Soviet winter wheat crop of 1971–72.[83]

These events mobilized national governments and the United Nations into action, and in 1974 they established the International Board for Plant Genetic Resources (now known as the International Plant Genetic Resources Institute, or IPGRI), which cobbled together a global network of gene banks. The network includes university breeding programs, national government seed storage units, and, most prominently, the Consultative Group on International Agricultural Research (CGIAR), a worldwide network of 16 agricultural research centers originally established to bring the Green Revolution to developing countries, and funded primarily by the World Bank and international aid agencies.[84]

The number of unique seed samples, or "accessions," in gene banks now exceeds 6 million. The largest chunk of these accessions, more than 500,000, are in the gene banks of CGIAR centers such as the International Rice Research Institute (IRRI) in the Philippines (the world's premier rice-breeding institute) and the International Wheat and Maize Improvement Center (CIMMYT) in Mexico. At least 90 percent of all gene bank accessions are of food and commodity plants, especially the world's most intensively bred and economically valuable crops. (See Table 5.) By the late 1980s, IPGRI regarded a number of these crops, such as wheat and corn, as essentially completely collected; that is, nearly all of the known landraces and varieties of the crop are already represented in gene banks. Others have questioned this assessment, however, arguing that the lack of quantitative studies of crop gene pools makes it difficult to ascertain whether even the best-studied crops have been adequately sampled.[85]

There are additional reasons for interpreting gene bank totals conservatively. The total annual cost of maintaining all accessions currently in gene banks is about $300 million, and some facilities, hard-pressed for operating funds, cannot maintain their collections under optimal physical conditions. Seeds that are improperly dried or kept at room temperature rather than in cold storage may begin to lose viability within a few years. At this point, they must be "grown out"—germinated, planted, raised to maturity, and then reharvested, which is a time-consuming and labor-intensive activity when repeated for thousands of accessions per year. These problems suggest that an unknown portion of accessions is probably of questionable viability. A recent project by the U.S. Maize Crop Germplasm Committee and other institutions to evaluate and regenerate 30,000 distinct accessions of corn collected from Latin America found that more than half had lost significant viability to the point where they were not usable for breeding purposes. Moreover, the majority of Latin American corn accessions have yet to be evaluated, and recent proposals for funding to extend the regeneration work have been rejected by the U.S. Agricultural Research Service.[86]

Only 13 percent of gene-banked seeds are in well-run facilities with long-term storage capability, and even the crown jewels of the system, such as the U.S. National Seed

TABLE 5

Gene Bank Collections for Selected Crops

Crop	Accessions in Gene Banks	Estimated Share of Landraces Collected
	(number)	(percent)
Wheat	850,000	90
Rice	420,000	90
Corn	262,000	95
Sorghum	168,500	80
Soybeans	176,000	70
Common Beans	268,500	50
Potatoes	31,000	80–90
Cassava	28,000	35
Tomatoes	77,500	90
Squashes, Cucumbers, Gourds	30,000	50
Onions, Garlic	25,000	70
Sugarcane	27,500	70
Cotton	48,500	75

Source: See endnote 85.

Storage Laboratory, have at times had problems maintaining seed viability rates. For extensively gene-banked crops (primarily major grains and legumes) where large collections are duplicated in different facilities, the odds of losing the diversity already on deposit are reduced. But for sparsely collected crops whose accessions are stored at just one or two sites, the possibility of genetic erosion remains disquietingly high.[87]

Despite such drawbacks, off-site facilities remain indispensable for conservation. In some cases, botanical gardens and gene banks have rescued species whose wild populations are now gone. One such species, the giant fan palm *Corypha taliera*, is thought to be extinct in the wild in its native range in West Bengal, India, due to indiscriminate felling and habitat loss. Fortunately, individuals of this spectacular palm are cultivated in botanical gardens in India and the United

States, and there is hope that a reintroduction program can be mounted.[88]

Gene banks are proving their worth as well in returning traditional crop varieties to farmers following natural and man-made disasters. Although the uplands of East Africa are not the center of domestication for common beans, the farmers of the region adopted them as their own several centuries ago, and have developed the world's richest mix of local bean varieties. When Rwanda was overwhelmed by civil conflict in 1994, the height of the genocidal violence occurred during the February-to-June growing season, greatly reducing harvests and raising the prospect of widespread famine. Amid the relief contributions that flowed into the country once the situation had stabilized were stocks of at least 170 bean varieties that had been previously collected in Rwanda and stored in gene banks worldwide. These supplies helped ensure that Rwandan farmers had stocks of high-quality, locally adapted beans for planting in the subsequent growing season. The need for international relief agencies to have supplies of locally adapted seed was similarly evident in the aftermath of the nearly decade-long civil conflict in the West African nations of Sierra Leone and Liberia, where local seed supply networks were extensively disrupted and national gene bank facilities were severely damaged.[89]

Still, even the most enthusiastic boosters of botanical gardens and gene banks recognize that such facilities, even when impeccably maintained, provide only one piece in the conservation puzzle. Off-site storage takes species out of their natural ecological and cultural settings. The seeds of a wild tomato can be sealed in a glass jar and frozen for safekeeping, but left out of the cold are the plant's pollinators, its dispersers, and all the other organisms and relationships that have shaped the plant's unique evolution. When stored seed populations are grown out over several generations off site, in time they can even lose their native adaptations and evolve to fit instead the conditions of their captivity.[90]

For these and other reasons, plant diversity can be maintained only by protecting the native habitats and ecosystems where plant species have evolved. Countries have safeguarded habitat primarily by establishing national parks, forest reserves, and other formally protected areas. During this century, governments have steadily increased protected area networks, and they now encompass nearly 12 million square kilometers, or about 8 percent of the Earth's land surface. Many protected areas guard irreplaceable botanical resources; Malaysia's Mount Kinabalu National Park, for instance, contains the unique vegetation of Southeast Asia's highest peak. A few reserves have even been established specifically to protect plant genetic resources, among them a forest reserve established by the Indian government to safeguard several wild species of citrus trees, and Israel's Ammiad reserve, near the Sea of Galilee, which protects grassy hillsides of wild emmer, one of the ancestors of wheat.[91]

Yet current protected area networks also have major limitations as far as protecting plants is concerned. Many parks are established at sites of surpassing scenic beauty or historical importance, but are not particularly significant for biodiversity. Meanwhile, many rare plant populations and highly diverse plant communities such as tropical deciduous forests are greatly underprotected. In addition, protected areas officially decreed on paper often are, in reality, minimally implemented by chronically underfunded and understaffed natural resource agencies. But perhaps the most fundamental limitation of national parks, wilderness areas, and similarly strict designations arises when they conflict with the cultural and economic importance that plants and other natural resources hold for local communities.[92]

Much of the natural wealth that conservationists have sought to protect is actually on lands and under waters long managed by local people. Indigenous societies worldwide have traditionally protected prominent landscape features like mountains or forests, designating them as sacred sites and ceremonial centers. In heavily settled parts of West Africa, sacred groves hold some of the last remaining populations of important medicinal plants. On Samoa and other Pacific islands, communities manage forests to produce wild

foods and medicines, raw materials for canoes and household goods, and other benefits. In industrialized countries,
too, local landowners are an important factor in the conservation equation. In the United States, for instance, a significant proportion of the rarest plant species (perhaps as much
as 50 percent) are found on relatively small private land-
holdings, not on existing conservation lands or large prop-
erties more easily acquired for conservation.[93]

Not surprisingly, actions such as evicting long-term residents from newly designated forest reserves, or denying
them access to previously harvested stands of useful plants,
have generated a great deal of ill will toward protected areas
worldwide. In certain cases where long-term residents have
been made equal partners in managing protected lands,
however, workable alternatives are emerging. In the Indian
state of West Bengal, 320,000 hectares of semi-deciduous sal
forest is managed jointly by villagers and the state forestry
department, with villagers taking primary responsibility for
patrolling nearby forest stands, in exchange for the right to
collect certain forest products. Since joint management
began in 1972, the status of the sal forests has improved, and
regenerating stands now provide villagers with medicines,
firewood, and wild-gathered foodstuffs.[94]

Medicinal plants also feature prominently in a 2,400
hectare rainforest reserve in the Central American country of
Belize. The reserve, which is government owned but managed by the Belize Association of Traditional Healers, was
established in 1993 as an extractive reserve for medicinal
plants commonly used by local expert healers. Yet even with
considerable local support, the future of the Belize reserve is
still not secure. Logging and agricultural development are
increasing in adjacent areas, and a slackening of political support for the reserve on the part of local or national government officials could jeopardize the work under way there.[95]

Belize's experience is a familiar one. Unless we can
develop a new balance between human needs and those of
the natural world, no park or reserve, no matter how well
intentioned or designed, will be able to head off further dete-

rioration of plant biodiversity. Protecting the habitat of endangered plants with a nature reserve or funding the activity of a gene bank is but the start of wise stewardship of biodiversity. Stewardship also means making sure that the patterns by which we use plants are sustainable and appropriate. It means revitalizing diversity in agricultural landscapes so that we grow food and fibers in more efficient, stable, and healthful ways. And most fundamentally, stewardship depends on an informed and dedicated citizenry willing to lead in conserving biodiversity when governments and other institutions delay or decline to take action.

Bringing Diversity Back Home

The best map we have for developing a society more friendly to plant diversity is contained in the Convention on Biological Diversity (CBD), which came out of the 1992 Rio Earth Summit. Now ratified by 172 countries, the CBD legally binds signatory governments to reversing the ongoing global decline not just in plants but in all biological diversity. Under the convention, countries are supposed to accomplish this feat by implementing protected area systems and off-site conservation facilities; by making certain that biological resources are used in ecologically sustainable ways; and by ensuring that the benefits of biodiversity—such as new medicines—are shared fairly and equitably. To implement the CBD, countries are required to take action on a number of fronts. (See Table 6.)[96]

The CBD is by far the most comprehensive environmental treaty ever negotiated. Not surprisingly, its implementation has proceeded slowly, reflecting the ambivalence of governments about how strong they actually want the convention to be. The United States Congress, for one, has steadfastly refused to ratify the treaty. The CBD has served to channel international funding donated by countries in support of biodiversity conservation. This funding totalled some

TABLE 6

Requirements for Signatory Nations to the Convention on Biological Diversity

• Adopt national biodiversity strategies and action plans.

• Establish nation-wide systems of protected areas.

• Adopt measures that provide incentives to promote conservation and sustainable use of biological resources.

• Restore degraded habitats.

• Conserve threatened species and ecosystems.

• Minimize or avoid adverse impacts on biodiversity from the use of biological resources.

• Respect, preserve, and maintain knowledge, innovations, and practices of local and indigenous communities.

• Ensure the safe use and application of biotechnology products.

Source: See endnote 96.

$1.25 billion for 156 separate projects between 1991 and 1997, which were administered through an interim funding mechanism, the Global Environment Facility (GEF), established in 1991 as a joint initiative of the United Nations Development Programme (UNDP), the United Nations Environmental Programme (UNEP), and the World Bank. The GEF's effectiveness in administering these funds, however, has been questioned in several outside evaluations, including a 1997 report by a team of independent environmental consultants, who found that many GEF projects were poorly prioritized and not well executed, making at best a marginal contribution to achieving the CBD's objectives. Still, the CBD has been useful in serving as a regular, global forum for debating issues related to using and conserving biodiversity. Just simply maintaining the convention requires extensive dialogue between government representatives and non-governmental participants, including both activists and scientists.[97]

In the meantime, the challenge of turning the CBD's objectives into tangible progress is being taken up in countries and communities by progressive government programs, innovative grassroots organizations, and dynamic individual citizens. The breadth and variety of their work is far too great to cover comprehensively here, but one can get a sense of the complexity involved by looking specifically at two areas: how to better manage wild plant harvests and how to rejuvenate biodiversity in agricultural systems.

One of the seemingly most daunting objectives of the Convention is to promote the sustainable harvest of biological resources. There is an inherent tension at work when trying to conserve a plant by giving it value in a setting where many people earn their living by extracting value from nature. If social codes and customs are not already in place to control and regulate that extraction, the result of adding value is likely to be overharvest rather than sustainable use.

This tension helps explain why very few commercially marketed wild plants—be they plants yielding medicines, tropical timbers, or sweet-smelling extracts for perfumes—are harvested in ways or at levels that are sustainable. With economic globalization, it is increasingly easy to link consumer demand in distant cities and other nations with stocks of timber, medicinal herbs, and natural resources in faraway biodiversity-rich regions. Once lucrative markets emerge for a plant product, it can be very difficult for local communities alone to regulate the harvest through customary practices, particularly when politically and economically powerful outsiders seek a share of the resource bonanza. In many countries, governments are likewise unable to assert regulatory control, in large part because the agencies responsible for managing natural resources are often poorly funded, politically marginalized within the government bureaucracy, or heavily oriented toward a few influential natural resource uses, such as timber harvesting or livestock ranching.[98]

One case where diverse uses of wild plants are losing out is in the Indonesian provinces of Kalimantan (Indonesia's portion of the island of Borneo), which contains some of the largest remaining tracts of tropical forest in Asia. Although thinly populated by Asia's standards, the forests of

Kalimantan are far from empty. Resident Dayak peoples, among others, manage complex mixtures of swidden cultivation plots, rattan groves, forest gardens of fruit and rubber trees, and mature rainforest in which they gather a wide variety of plant and animal resources. In most areas, such diverse subsistence traditions have maintained an intact forest base—too successfully so, as it turns out, since the Indonesian government views forest lands as property of the state. Rather than recognizing local stewardship and supporting customary land claims, the government has granted extensive concessions for timber and, increasingly, for oil palm and pulpwood plantations, to companies controlled by Indonesia's economic elite. Although plantations are supposed to be targeted for lands already degraded by logging, the sheer size of the concessions (tens and even hundreds of thousands of hectares) means that many overlap with the farms and forest gardens of local communities. Villages that resist the concessionaires' usurpment of local lands have often seen their gardens bulldozed and their complaints to local government representatives ignored.[99]

The massive fires that consumed Kalimantan during 1997 and 1998 were started deliberately—mainly by plantation concessionaires, who took advantage of the unusually dry El Niño conditions to clear land quickly and evict local residents, and by immigrant farmers clearing land as part of government-sponsored transmigration projects. Although Dayak farmers, too, use fire to clear swidden plots, they do so skillfully and carefully—all the more so in drought years, when the odds of fire escaping into adjacent forest gardens are heightened. Between the drought and the fires, villagers' staple rice harvests in many parts of Kalimantan in 1997–98 were only a fraction of their normal levels. When the rice crop failed in past years, villagers could gather wild foodstuffs from the forest, or else sell rattan or rubber from their forest gardens and buy rice with the cash earned. This time, with their forests increasingly taken over by outsiders, thousands of villagers headed to towns for relief handouts, or sold their land to concessionaires, one-half hectare for a month's supply of rice.[100]

Not all situations are as bleak as Kalimantan's. There are bright spots where communities that commercially harvest wild plants have made substantial progress toward sustainability. In Mexico, ancient cone-bearing plants called cycads have been heavily exploited for their ornamental value, both for sale domestically and for export to the warmer regions of the United States, Japan, and Europe. Most cycads are harvested from the wild by uprooting or cutting, but the Francisco Javier Clavijero Botanical Garden in the state of Veracruz is working with local villages to reduce pressures on several overexploited species. In one community, Monte Oscuro, residents set aside a communal plot of dry forest to protect a relict population of cycads, in exchange for help with building a community plant nursery. Seeds are collected from the wild plants, then germinated and tended in the nursery by villagers who have received training in basic cycad propagation from the botanical garden staff. Some of the young cycads are returned to the forest to offset any potential downturn in the wild population from the seed harvest. The rest are sold and the profits deposited in a community fund.[101]

The experience gained from such collaborative projects illuminates some of the foundations of sustainable harvests. To start with, a system clearly establishing locally and nationally recognized tenure over the harvested resource is essential—either in the form of rights to collect a plant product, or ownership of the land it grows on. At Monte Oscuro, the cycads grow on land collectively owned by the community, a legally recognized category of land ownership in Mexico called an *ejído*, which establishes that no individual person has a right to harvest the plants or the seeds without the consent of the community. Another important factor is that Monte Oscuro residents have other options for making a living (small-scale agriculture and livestock husbandry, for example) besides cycad harvesting, and therefore can afford to forego the immediate monetary benefits of harvesting the adult plants in exchange for delayed future benefits from the nursery production. A third key asset is understanding the ecology and management of the type of plant being harvested. At Monte Oscuro, local residents and professional botanists have combined their knowledge about cycads to figure out the best way to rejuvenate the local wild stand and propagate young plants in a nursery setting.[102]

At the same time, making a harvest sustainable involves issues and trends that extend well beyond village-level economics and social dynamics. The biggest hurdle facing the Veracruz cycad project is finding good markets for the young plants the communities are producing; cycads are slow growing and horticultural buyers prefer larger, older plants. Since most large cycads that enter the ornamental trade are wild-harvested, better monitoring and enforcement of the international cycad trade might increase the marketability of nursery-raised plants. About half of Mexico's cycad species are listed with CITES, which provides a powerful legal tool for controlling international traffic in threatened plants and animals. CITES, established in 1973, is generally regarded as one of the more effective international environmental treaties. It prohibits trade in the most highly endangered species (listed in the Treaty's Appendix I), and keeps watch on vulnerable species (listed in Appendix II) by requiring that countries issue a limited number of permits for the species' export and import between signatory countries. Although CITES provides powerful tools for enforcing sustainable harvests, its effectiveness still depends on a country's ability to set permit levels that reflect the species' status in the wild, and to stop illegal trade that goes on outside the permit system.[103]

Another option for cycads and other heavily harvested plants is establishing programs to certify that plants entering into trade are coming from "environmentally friendly" sources that meet high ecological and social standards. Environmental certification programs have already been established for bananas, coffee, chocolate, oranges, and, most prominently, temperate and tropical wood products. By mid-1998, nearly 10 million hectares of forest worldwide had been certified as being under ecologically sound timber management, primarily in Northern Europe and North America. This acreage is still only a small fraction of the world's forests under timber harvest, particularly for tropical regions, but the number of certified timber-producing operations is expected to continue to grow rapidly. A good deal of the credit for the growth of certified timber operations goes to the Forest Stewardship Council, a coalition established in 1993 by environmental groups, progressive foresters and businesses, and indigenous peoples' representatives.[104]

To guide action on sustaining plant diversity in agriculture, the signatory parties to the CBD have endorsed a global Plan of Action for conserving plant genetic resources, negotiated in Leipzig, Germany, in June 1996 at a technical conference sponsored by the FAO. It spells out priority actions for governments and agencies in four main areas—increasing support for off-site conservation of genetic resources in gene banks and other facilities; helping farmers maintain, restore, and improve the diversity of crops they cultivate on their farms; improving the utilization of plant genetic resources by promoting sustainable agriculture and developing new markets for lesser-known crops; and building the capacity of government agencies and other institutions to conserve, monitor, and educate their citizenry about the value of agricultural biodiversity. Comprehensive financing of the Action Plan would require approximately $455 million annually, but donor countries at Leipzig did not promise any specific new funds, leaving FAO and other agencies responsible for monitoring the Plan to urge its implementation through existing national and international programs.[105]

In some communities in Zimbabwe, villagers contribute seeds annually to a communal seed stock.

Although the Action Plan is already guiding the work of the FAO and CGIAR centers, its ultimate success, like that of the CBD, depends on how well it can be carried out in national and local arenas. Improving conservation as we traditionally conceive of it—gene banks, protected areas—is an important goal of the Action Plan. In addition, it essentially advocates agricultural development that strengthens rather than simplifies plant diversity to meet the needs and goals of all farmers—from subsistence-oriented farmers who still maintain diverse agricultural landscapes, to farmers who are firmly integrated into the commercial marketplace.[106]

Helping farmers maintain and improve the diversity of the crops they grow requires understanding the particular cultural, esthetic, economic, and agronomic reasons why farmers maintain elements of traditional farming, such as unique crop variety mixtures. For instance, the Hopi, a Native American people of the southwest United States, continue to raise indigenous corn and lima bean varieties not only because the plants are well adapted to the sandy soils the Hopi cultivate, but also because the germinating seeds are indispensable for their religious ceremonies. Mende farmers in Sierra Leone continue to grow native red-hulled African rice for a similar reason (most rice from Asia is brown-hulled and unsuitable for certain Mende rituals). Andean peasant farmers still grow pink and purple potatoes, big-seed corn, quinoa, and other traditional crops because that is what they themselves prefer to eat; they grow commercial varieties strictly for cash income.[107]

One optional strategy to help farmers maintain crop diversity involves supporting farmers' informal networks of seed exchange and procurement, so as to improve their access to diverse seed sources. In some rural communities in Zimbabwe, villagers contribute seeds annually to a communal seed stock. At the start of the planting season, the seeds are redistributed to all community members, a step that gives villagers access to the full range of varietal diversity present in the immediate vicinity and ensures that no one goes without seeds for planting. Grassroots organizations in Ethiopia, Peru, Tonga, and many other countries have sponsored community seed banks, regional agricultural fairs, seed collection tours, demonstration gardens, and similar projects to promote informal seed exchange between farmers, increase their access to crop diversity, and help them replenish seed stocks after poor harvests.[108]

Grassroots organizations and innovative national plant breeding programs in developing countries have pioneered another promising approach to strengthen diversity in farmers' fields: reorienting formal plant breeding (like that practiced at national and international research stations and private companies) to better address the specific local needs of farmers. Typically, plant breeders create uniform, widely adaptable "pure-bred" varietal lines, but only toward the end of the process are the lines evaluated with farmers. Participatory plant breeding methods, on the other hand, involve farmers at all stages. In the most advanced participatory approaches, breeders and farming communities work together in farmers' fields rather than at research stations to evaluate, select, combine, and improve a wide range of varieties, both those available locally and those from other regions. In this way, participatory plant breeding can improve the suite of locally adapted varieties without resorting to varietal uniformity; this approach maintains—or potentially even enhances—the genetic diversity present in farmers' fields.[109]

Grassroots approaches are helping conserve crop biodiversity in industrialized countries as well. In the United States, a nonprofit organization called the Seed Saver's Exchange coordinates a network of 8,000 farmers and backyard gardeners who each year grow and exchange about 15,000 heirloom and open-pollinated (or non-hybrid) varieties of fruits, vegetables, flowers, and herbs that have been passed down from generation to generation within families. Since the Exchange was founded in 1975, members have distributed over 500,000 samples of seeds not available in commercial seed catalogs. Similar efforts to link people who grow heirloom crops are under way in Australia and Europe.[110]

Another U.S.-based organization called Native Seeds/SEARCH takes a regional focus, working to preserve traditional crops and farming methods in the southwestern United States and northern Mexico. In addition to maintaining a seed bank of regional landraces, Native Seeds documents indigenous knowledge about crops and agricultural practices, and sponsors numerous local festivals and events featuring traditional foods of the region. Native Seeds has also worked closely with Native American peoples of the southwestern United States, who suffer some of the world's highest rates of adult-onset diabetes (up to 35 percent of the population in some groups), a trend thought to be related to shifts in diet with acculturation. Native Seeds has sponsored research demonstrating how the high concentrations of soluble fiber, tannins, and other compounds in traditional foods help control the onset of diabetes, and has conducted outreach with Native American health professionals on incorporating native foodstuffs like mesquite flour and cactus fruits into a diabetic diet.[111]

It is no accident that organizations and efforts begun by small groups of dedicated, dynamic, and focused people have led the way in conserving and reinvigorating diversity in agriculture. Implementing the FAO Action Plan, however, will require taking the knowledge and experience gained from work at the grassroots level, and applying it to modify the world's predominant mode of producing and distributing food and fiber: commercially oriented farms and plantations, where crops are grown primarily in monocultures. This agro-industrial model is heavily favored by agribusiness and governments and has spawned a global food distribution and processing industry.

Agro-industrial food production has had spectacular short-term success in raising agricultural yields, but has worked against genetic diversity in crops, ecological diversity in farming landscapes, and cultural knowledge about biodiversity within farming communities and societies. One reason for this failure is that agro-industrial food production tends to favor centralization and economies of scale rather than localized, independent variability. Participatory plant breeding, for instance, tends to produce benefits that are spread diffusely among farmers, and are not so easily appropriated for commercial gain—which helps explain why commercial seed suppliers have not led the way in developing participatory breeding approaches.[112]

The CGIAR centers are somewhere in the middle. Their mandate is specifically to improve the food security of the poor in developing countries, and they are actively exploring participatory approaches, but they also remain heavily involved in standard breeding programs. For instance, the corn and wheat center CIMMYT recently collaborated with university breeders and seed companies in Mexico to develop better-yielding corn varieties targeted for highland areas of the country where corn landraces continue to be grown by small-scale farmers under diverse environmental conditions. The collaborators chose to focus on hybrid corn varieties which, if tailored to the environmental and economic constraints facing highland Mexican farmers, could boost crop yields and farm productivity. But farmers will lose the option of saving their seeds and adapting them further to their particular local conditions. In adopting hybrid corn varieties, farmers are implicitly trusting that seed companies and international institutions will provide seeds suited to their particular environmental, economic, and social requirements— both present and future. The fact that a substantial percentage of the corn crop in Mexico and other Latin American countries is still planted to non-hybrid (or "open-pollinated") varieties suggests that many subsistence farmers still prefer to keep responsibility for seed supplies in their own hands.[113]

Another reason why biodiversity has declined under prevailing agricultural practices is that governments have become tightly wedded to the benefits of agro-industrial production, and thus actively promote it with their policies. Farmers in most southern African countries, for instance, are only eligible to participate in government agricultural credit programs if they agree to plant improved, professionally bred varieties of crops. Of course, governments often have noble motives for these policies, including increasing national self-sufficiency in food production, but broad-brush approaches like these can easily backfire. When such policies appear likely to lead to loss of biodiversity, they deserve particularly careful scrutiny and comparison with alternatives, to make sure the benefits are worth the costs involved—over the long as well as the short term.[114]

One initiative that could have benefited from greater scrutiny at its onset is the promotion by the U.S. Agency for International Development and the World Bank of non-traditional commercial crops throughout Latin America for export to high-value markets, primarily in the United States. These crops include asparagus, snow peas, berries, melons, and cut flowers, which unlike bananas or coffee did not have long-established export markets and which often were not grown previously for local consumption. While non-traditional crops have diversified exports and boosted foreign exchange earnings, benefits at the farm and community level have been more muted. And a number of harmful side-effects have appeared. Non-traditional export crops are typically grown in unrotated monocultures with heavy doses of agrochemicals, creating significant health risks for farmworkers and often replacing diverse indigenous crop mixtures that farmers had previously planted for subsistence needs. In some cases, export-oriented crops have exacerbated environmental degradation—as in Guatemala, where non-traditional vegetable production expanded onto steep hillsides and displaced subsistence cultivation into previously forested areas. Moreover, the economic benefits generated by new exports have tended to be captured by larger farms and commercial operations rather than small producers—except in cases where export buyers and producer cooperatives have specifically linked their activities to generating benefits for small-scale farmers.[115]

However badly needed, a major policy shift in support of agricultural biodiversity is not likely to occur without profound changes in peoples' awareness of plant biodiversity's benefits, their desire to change existing practices, and their willingness to try new approaches. On this point there is cause for cautious optimism: biodiversity is increasingly viewed by people in the mainstream of agricultural production as a resource that can improve the stability and sustainability of modern agriculture, and whose rejuvenation makes both good ecological sense and, in the long run, good economic sense as well.

In the state of Iowa, the U.S.'s second leading agricultural state, a group called Practical Farmers of Iowa is collaborating with researchers at Iowa State University to identify agronomic practices that potentially both enhance diversity on farms and save money: alternative crop rotations, planting cover crops to control weeds and boost soil fertility, and maintaining non-crop plants along field margins to provide habitat for beneficial insects, for instance. With Iowa's two leading crops, corn and soybeans, selling in fall 1998 for less than it cost farmers to produce them, there is unprecedented interest among farmers in using diversity-enhancing methods to help reduce their farming costs.[116]

There is unprecedented interest among farmers in using diversity-enhancing methods to reduce their farming costs.

At the international level, crop-breeding programs within CGIAR institutions are making wiser use of genetic diversity than they did during previous decades. During the early 1970s, IRRI released finished rice varieties directly, resulting in a handful of high-yielding rice varieties predominating across South and Southeast Asian rice paddies. By the 1990s, however, IRRI had switched to emphasizing unfinished breeding lines of rice that were distributed to different national rice breeding programs, which then produced finished rice varieties specifically adapted to the prevailing conditions in their country. This policy is one reason why the average number of traditional rice varieties contributing to the genetic composition of newly released varieties has risen from about three in the 1960s to about eight today. Professionally bred wheat varieties released in developing countries also show a trend of increasingly diverse pedigrees, and data from India suggest that wheat fields in that country's breadbasket have become more genetically diverse as the

Green Revolution has matured.[117]

On the other end of the production system, consumers who support diverse ways of producing food have a key contribution to make. The global market for organically grown crops has expanded by 20 percent annually during the 1990s, and this strong consumer demand has enabled thousands of farmers to switch to cultivation methods that make greater use of biodiversity for controlling pests and weeds, maintaining soil fertility, and other agronomic benefits. Farmers' markets, where farmers sell fruits and vegetables directly to consumers, are popular venues for marketing many crop varieties that do not meet the narrow standards of commercial food processors and distributors. In addition, a network of over 70,000 environmentally conscious cooks, restauranteurs, and gastronomes in Europe and the U.S. are working to revitalize regional food specialties and culinary traditions drawing on local foodstuffs, in what is being termed a "slow-food" movement.[118]

As encouraging as these trends are, the scale of what remains to be accomplished to reverse the erosion of plant biodiversity, both in agriculture and in wildlands, presents a sobering challenge. While many people are pursuing alternatives to unsustainable patterns in agriculture and natural resource use, others remain heavily invested in continuing them. Despite growing public demand for organic foods, for instance, less than one tenth of one percent of the research projects funded by the U.S. Department of Agriculture in 1995 and 1996 were focused on organic cultivation methods. Without action to address the underlying social and economic trends eroding biodiversity, progress on one front like sustainable agriculture can easily be offset by negative developments in a related arena. How well we conserve plant biodiversity and other natural resources depends ultimately on what kind of society we shape around them.[119]

Sharing the Benefits and Obligations

Governments can begin to chart a new social course by resolving a prominent policy issue affecting plant diversity today: how to distribute biodiversity's economic benefits fairly and equitably. Establishing a system of intellectual property rights to plant resources and knowledge about them has proved contentious because of a fundamental pattern—plant diversity (both wild and cultivated) is maintained mostly by developing countries and indigenous peoples, but the economic benefits it generates are disproportionately captured by industrialized nations. For most of this century, plant diversity has been treated as the "common heritage" of humankind, freely available to anyone who can collect it and use it, with proprietary ownership only granted, via patent or variety protection law, to individuals who demonstrate that they possess trade secrets about a plant or have uniquely improved a cultivated variety. Without the principle of common heritage, it is unlikely that agricultural research programs could have realized the genetic improvements—and thus the astounding productivity gains—in wheat, rice, and other staple crops this century. Nor is it likely we would have the full range of drugs and biomedical compounds available today for treating illness and disease.[120]

Yet common heritage does not mean that people and industries who use crop varieties, medicinal plants, and other biological resources have no obligation to those who provide them. Since the early 1980s, the question of what exactly this obligation should entail has been hotly debated. Some people argue that the obligations of common heritage are fulfilled adequately by the return flow of non-proprietary medicines and improved crop varieties to developing countries, and by the research and training support that international institutions have given developing nations for using plant diversity. Others claim that these benefits do not match those gained by industrial users of biodiversity, or argue that common heritage needs to be re-thought so that its benefits reach more effectively the indigenous people and traditional farmers who have generated and maintained detailed knowledge about biodiversity.[121]

It is now widely recognized that common heritage entails a common responsibility for conservation. One way proposed to acknowledge this responsibility is to establish an international fund supporting continued community-based management and conservation of plant resources. Such a step appears the most practical means of compensation for the large amount of plant biodiversity that is already in the public domain (such as the millions of seed accessions in gene banks or plants widely used as herbal medicines), since establishing exactly who deserves compensation for commercial innovations from these plant resources would be a Herculean task.[122]

Other proposals for redressing the limitations of common heritage center on expanding proprietary claims to plant resources. Grassroots advocates argue, for example, that indigenous people deserve "traditional resource rights" based on demonstrating unique cultural knowledge about or long-standing use of a biological resource. They advocate giving such rights international legal standing equivalent to that afforded to patent rights. Recognition of such rights will require, at a minimum, negotiating equitable benefit-sharing agreements at the community level whenever plants or indigenous knowledge about them is collected by researchers.[123]

To date, formal agreements for sharing the benefits of plant diversity have been negotiated most extensively in the search for new pharmaceuticals from plants in biodiversity-rich developing countries. The first such "bioprospecting" agreement was announced in 1991 between Merck Pharmaceuticals and Costa Rica's National Institute of Biodiversity (InBio), in which Merck paid InBio $1.1 million for access to plant and insect samples and promised to share an undisclosed percentage of royalties from any commercial products that resulted.[124]

There are now at least a dozen bioprospecting agreements in place worldwide involving national governments,
indigenous communities, conservation groups, start-up
companies, and established corporate giants. These agreements have proliferated in the wake of the 1992 Convention
on Biological Diversity, which recognized that countries
have sovereignty over plants and other genetic resources
within their borders. Most legitimate bioprospecting agreements have followed the Merck-InBio model, with a modest
up-front payment and a promise to
return between one quarter of 1 percent **Critics suspect**
and 3 percent (depending on the project) **that not all**
of any future royalties to the biodiversity **bioprospecting**
holders. Bioprospecting proponents
argue that with the huge cost ($200– **agreements**
$350 million) of bringing a new drug to **are negotiated**
market, companies cannot afford to **on an equal**
share a higher percentage of royalties. **footing.**
Critics, however, suspect that many
bilateral bioprospecting agreements are
not negotiated on an equal footing, and
more powerful counterparts are able to win better deals.
When a biotechnology firm approached the U.S. government about prospecting for unique microbes inhabiting the
geysers and hot springs of Yellowstone National Park, for
instance, the Park Service negotiated a royalty share of 10
percent. Moreover, not all bioprospecting agreements auto-
matically uphold traditional resource rights; many have
been negotiated at a national rather than a community level,
involving governments that many indigenous people think
do not adequately represent—that indeed sometimes actively undermine—their interests.[125]

In contrast with bioprospecting, resolving who owns
the world's crop genetic resources is being negotiated multi-
laterally, in factious diplomatic arenas. In 1989, the FAO
adopted a Farmers' Rights proposal that would compensate
farmers for their contribution to biodiversity via an international trust fund to support the conservation of plant genet-

ic resources. The CBD also called for incorporating Farmers' Rights subsequent to further international negotiations. There has been no official endorsement of this concept, however, from the industrial nations who would provide the compensation, and the fund has remained unimplemented. During the most recent round of international negotiations in June 1998, the European Union appeared ready to support Farmers' Rights, but Australian, U.S., and Canadian diplomats consistently refused to address the issue.[126]

Meanwhile, the intellectual property agenda of industrial countries is being advanced by the WTO. All countries acceding to the WTO's General Agreement on Tariffs and Trades are required to establish a system for protecting breeders' rights through plant variety patents. They can either adopt the system of administering patents and breeders' rights followed by industrial nations under the International Union for the Protection of New Varieties of Plants (UPOV), or instead design their own unique system. The UPOV Convention was originally established in 1968, substantially expanded in 1978, and then revised again in 1991. Initially the UPOV gave farmers the right to save commercial seed for their own use, but the 1991 version allowed signatory countries to revoke this right. Some countries, including India, are looking at structuring their own plant patent systems to also acknowledge farmers' rights, but it is unclear whether the WTO will approve such arrangements.[127]

Despite the foot-dragging in international arenas, de facto boundaries are emerging for what will and will not be tolerated in the expropriation of crop genetic resources from a public to a private good. In May 1997, two Australian agricultural centers applied for proprietary breeders' rights on two varieties of chickpeas. Their application sparked an international uproar because the Australian breeders had obtained both varieties from a CGIAR gene bank, which had provided the seeds with the understanding they were to be used for research and not for direct financial gain. Moreover, the Australians did little breeding to improve the two chickpeas, one of which was a landrace widely grown by Indian farmers. Ultimately, the Australian government bowed to international pressure and rejected the patent application. The CGIAR, for its part, subsequently called for a moratorium on all claims for proprietary breeding rights involving germplasm held in trust by CGIAR-sponsored gene banks.[128]

In the struggle to establish fair and respectful ways of using plant diversity, it is easy to overlook our mutual dependence: from urban consumers, to peasant farmers, we all depend on wise stewardship of biodiversity. Throughout human history, we have drawn on plants for both spirit and subsistence, and in the end there is no separating the two. Cases where people inadvertently or intentionally deplete plant diversity are all too common. Yet as the thousands of crop varieties still cultivated by traditional farmers and backyard gardeners demonstrate, we are equally capable of fostering and enhancing biodiversity, given the proper means, incentives, and outlook.[129]

From reducing overexploitation of medicinal plants to codifying traditional resource rights for biodiversity stewards, many options are available for developing cultural, economic, and political links that support plant diversity rather than diminish it. Such steps are not just about meeting treaty obligations or decreeing new protected areas—they are part of a larger process of shaping ecologically literate civil societies that can live in balance with the natural world. And that process is something we can all contribute to.

Notes

1. Total area for transgenic varieties from Charles C. Mann, "Biotech Goes Wild," *Technology Review*, July/August 1999; total number of transgenic varieties introduced from Marian Burros, "Lines Redrawn in International Food Fight," *The New York Times*, 14 July 1999; industry and government view of biotechnology from speech by Robert B. Shapiro, CEO, Monsanto Corporation, before the Biotechnology Industry Organization, June 1998, <http://www.monsanto.com/monsanto/mediacenter/speeches/98jun17_sh apiro.html>, viewed 6 July 1999; and from comments by Dan Glickman, U.S. Secretary of Agriculture, in "Agricultural Biotechnology: An Important New Tool for the 21st Century," <http://www.monsanto.com/monsanto/ mediacenter/background/99apr27_promo4.html>, viewed 6 July 1999.

2. Reduced pesticide needs of transgenic potatoes and their status as most pesticide-dosed crop from Michael Pollan, "Playing God in the Garden," *The New York Times Magazine*, 25 October 1998; and from Monsanto Corporation, "New Leaf® Potatoes: Contributing to Sustainability," <http://www.monsanto.com/ag/articles/PlantBiotech/Potatoes.html>, viewed 6 July 1999.

3. Irish potato famine from Cary Fowler and Pat Mooney, *Shattering: Food, Politics, and the Loss of Genetic Diversity* (Tucson, AZ: University of Arizona Press, 1990); resurgent potato blight information from Kurt Kleiner, "Save Our Spuds," *New Scientist*, 30 May 1998; from Pat Roy Mooney, "The Parts of Life: Agricultural Biodiversity, Indigenous Knowledge and the Third System," *Development Dialogue*, nos. 1–2, 1996; from "Potatoes Blighted," *New Scientist*, 21 March 1998; and from International Potato Center, "An Opportunity for Disease Control," <http://www.cgiar.org/cip/blight/ lbcntrl.htm>, viewed 23 October 1998. Control of late blight in U.S. from Pollan, op. cit. note 2. Late blight is *Phytopthora infestans*.

4. Potato domestication in Andes and CIP collecting and breeding efforts from Z. Huamán, A. Golmirzaie, and W. Amoros, "The Potato," in Dominic Fuccillo, Linda Sears, and Paul Stapleton, eds., *Biodiversity in Trust: Conservation and Use of Plant Genetic Resources in CGIAR Centers* (Cambridge, U.K.: Cambridge University Press, 1997); potato varieties in traditional Andean villages from Miguel A. Altieri, *Agroecology: The Scientific Basis of Alternative Agriculture* (Boulder, CO: Westview Press, 1987); wild potato descriptions based on J.G. Hawkes, *The Potato: Evolution, Biodiversity and Genetic Resources* (Washington, DC: Smithsonian Institution Press, 1990).

5. Edward O. Wilson, *The Diversity of Life* (Cambridge, MA: Belknap Press, 1992).

6. Global human population from United Nations Population Division, *World Population Projections to 2150* (New York: U.N. Secretariat, 1998); use of Earth's biological systems from Peter M. Vitousek, Harold A. Mooney,

Jane Lubchenco, and Jerry M. Melillo, "Human Domination of Earth's Ecosystems," *Science*, 25 July 1997; estimated species extinction rates from Nigel Stork, "Measuring Global Biodiversity and Its Decline," in Marjorie L. Reaka-Kudla, Don E. Wilson, and Edward O. Wilson, eds., *Biodiversity II: Understanding and Protecting Our Biological Resources* (Washington, DC: Joseph Henry Press, 1997); total number of threatened species calculated from Jonathan Baillie and Brian Groombridge, eds., *1996 IUCN Red List of Threatened Animals* (Gland, Switzerland: World Conservation Union (IUCN), 1997); Kerry S. Walter and Harriet J. Gillett, eds., *1997 IUCN Red List of Threatened Plants* (Gland, Switzerland: IUCN, 1997).

7. Plant-based drugs from Norman R. Farnsworth, "Screening Plants for New Medicines," in E.O. Wilson, ed., *Biodiversity* (Washington, DC: National Academy Press, 1988).

8. Crop variety losses from Food and Agricultural Organization of the United Nations (FAO), *The State of the World's Plant Genetic Resources for Food and Agriculture* (Rome: 1996).

9. Chris Bright, "Tracking the Ecology of Climate Change," in Lester R. Brown, Christopher Flavin, and Hilary French, *State of the World 1997* (New York: W.W. Norton & Company, 1997).

10. Risks of genetic engineering from Mann, op. cit. note 1; and from Carol Kaesuk Yoon, "Pollen from Genetically Altered Corn Threatens Monarch Butterfly, Study Finds," *The New York Times*, 20 May 1999; EU moratorium from Debora MacKenzie, "Gene Crops Face New Hurdles," *New Scientist*, 3 July 1999.

11. Population in poverty from United Nations Development Program (UNDP), *Human Development Report 1997* (New York: Oxford University Press, 1997).

12. Robert E. Rhoades and Virginia D. Nazarea, "Local Management of Biodiversity in Traditional Agroecosystems," in Wanda W. Collins and Calvin O. Qualset, eds., *Biodiversity in Agroecosystems* (Boca Raton, FL: CRC Press, 1999).

13. Plant species total from Vernon H. Heywood and Stephen D. Davis, "Introduction," in S.D. Davis et al., eds., *Centers of Plant Diversity: Volume 3, The Americas* (Oxford, U.K.: World Wide Fund for Nature (WWFN), 1997) and from Anthony Huxley, *Green Inheritance* (London: Gaia Books Ltd., 1984).

14. Calculations of background extinction rates from David M. Raup, "A Kill Curve for Phanerozoic Marine Species," *Paleobiology*, vol. 17, no. 1 (1991). Raup's exact estimate is one species extinct every four years, based on a pool of 1 million species; Worldwatch translates this rate into one to 10 species per year based on current range of estimates of total species worldwide. Estimates for current rates of extinction are reviewed by Stork, op. cit. note 6; and by Stuart L. Pimm, Gareth J. Russell, John L. Gittleman, and Thomas M. Brooks, "The Future of Biodiversity," *Science,* 21 July 1995; to translate rates into whole numbers, we have assumed a total pool of 10 million species.

15. Wilson, op. cit. note 5.

16. At-risk figures from Walter and Gillett, op. cit. note 6; clover example from David Bramwell, "Top 50: A Millennial Flagship," *World Conservation,* June 1998.

17. Endemics on *Red List* from Walter and Gillett, op. cit. note 6; Hawaii endemism rate from S. D. Davis et al., *Plants in Danger: What Do We Know?* (Gland, Switzerland: IUCN, 1986); Hawaii endangerment from Brien Meilleur, "In Search of 'Keystone Societies'," in Nina L. Etkin, ed., *Eating on the Wild Side* (Tucson, AZ: University of Arizona Press, 1994).

18. Walter and Gillett, op. cit. note 6.

19. Colombian community endangerment from Andrew Henderson, Steven P. Churchill, and James L. Luteyn, "Neotropical Plant Diversity," *Nature,* 2 May 1991; from D. Olson et al., eds., *Identifying Gaps in Botanical Information for Biodiversity Conservation in Latin America and the Caribbean* (Washington, DC: World Wildlife Fund (WWF-U.S.), 1997); and from J. Orlando Rangel Ch., Petter D. Lowy C., and Mauricio Aguilar P., *Colombia: Diversidad Biótica II* (Santafé de Bogotá, Colombia: Instituto de Ciencias Naturales, Universidad Nacional de Colombia, 1997); western Australia from Wilson, op. cit. note 5; New Caledonia from J.M. Veillon, "Protection of Floristic Diversity in New Caledonia," *Biodiversity Letters,* vol. 1, nos. 3/4 (1993); longleaf pine forest from C. Kenneth Dodd, Jr., "Reptiles and Amphibians in the Endangered Longleaf Pine Ecosystem," <http://biology.usgs.gov/s+t/frame/d272.htm>, viewed 22 July 1999; and from S. Ware, C. Frost and P.D. Doerr, "Southern Mixed Hardwood Forest: The Former Longleaf Pine Forest," in W.H. Martin, S.G. Boyce, and A.C. Echternacht, eds., *Biodiversity of the Southeastern United States* (New York: John Wiley and Sons, 1993). Longleaf pine is *Pinus palustris.*

20. Southeast Florida habitats from Florida Natural Areas Inventory, county summaries, <http://www.fnai.org/dade-sum.htm> and <http://www.fnai.org/natcom.htm>, viewed 3 September 1998; from Lance Gunderson, "Vegetation of the Everglades: Determinants of Community Composition," in Steven M. Davis and John C. Ogden, eds., *Everglades: The Ecosystem and Its Restoration* (Delray Beach, FL: St. Lucie Press, 1994); and from *An Action Plan to Conserve the Native Plants of Florida* (St. Louis, MO: Center for Plant Conservation, 1995). Extent of invasive plants in U.S. from Randy G. Westbrooks, *Invasive Plants, Changing the Landscape of America* (Washington, DC: Federal Interagency Committee for the Management of Noxious and Exotic Weeds, 1998); Brazilian pepper is *Schinus terebinthifolius* and Australian casuarina is *Casuarina equisetifolia.*

21. Heywood and Davis, op. cit. note 13; Donald G. McNeil, Jr., "South Africa's New Environmentalism," *The New York Times,* 15 June 1998; Alan Hamilton, World Wide Fund for Nature-UK, letter to author, 30 June 1999.

22. Nitrogen overload effects from Anne Simon Moffatt, "Global Nitrogen Overload Problem Grows Critical," *Science,* 13 February 1998; climate change synergies from Bright, op. cit. note 9.

23. Oliver L. Phillips and Alwyn H. Gentry, "Increasing Turnover Through Time in Tropical Forests," *Science,* vol. 263, 18 February 1994.

24. CI study summarized in Lee Hannah et al., "A Preliminary Inventory of Human Disturbance of World Ecosystems," *Ambio,* July 1994; and in Lee Hannah, John L. Carr, and Ali Lankerani, "Human Disturbance and Natural Habitat: A Biome Level Analysis of a Global Data Set," *Biodiversity and Conservation,* vol. 4, 1995.

25. Edgar Anderson, *Plants, Man and Life* (Berkeley, CA: University of California Press, 1969); Carol J. Pierce Colfer, *Beyond Slash and Burn: Building on Indigenous Management of Borneo's Tropical Rain Forests* (Bronx, NY: New York Botanical Garden, 1997); John Schelhas and Russell Greenberg, eds., *Forest Patches in Tropical Landscapes* (Washington, DC: Island Press, 1996).

26. FAO, op. cit. note 8.

27. Total edible plant species from Timothy M. Swanson, David W. Pearce, and Raffaello Cervigni, *The Appropriation of the Benefits of Plant Genetic Resources for Agriculture: An Economic Analysis of the Alternative Mechanisms for Biodiversity Conservation,* Background Study Paper #1, FAO Commission on Plant Genetic Resources (Rome: FAO, 1994); date for span of agriculture from Bruce D. Smith, *The Emergence of Agriculture* (New York: Scientific American Library, 1995); hunter-gatherer plant use from Fowler and Mooney, op. cit. note 3. Wild grass harvesting from U.S. National Research Council (NRC), *Lost Crops of Africa: Volume I, Grains* (Washington, DC: National Academy Press, 1996); Thailand information from P. Somnasang, P. Rathakette, and S. Rathanapanya, "The Role of Natural Foods in Northeast Thailand," *RRA Research Reports* (Khon Khaen, Thailand: Khon Khaen University-Ford Foundation Rural Systems Research Project, 1988); Iquitos information from Rodolfo Vasquez and Alwyn H. Gentry, "Use and Misuse of Forest-Harvested Fruits in the Iquitos Area," *Conservation Biology,* vol. 3, no. 1 (January 1989); and Alwyn Gentry, "New Nontimber Forest Products from Western South America," in Mark Plotkin and Lisa Famolare, eds., *Sustainable Harvest and Marketing of Rain Forest Products* (Washington, DC: Island Press, 1992).

28. Maasai case from Timothy Johns, "Plant Constituents and the Nutrition and Health of Indigenous Peoples," in Virginia D. Nazarea, ed.,

Ethnoecology: Situated Knowledge/Located Lives (Tucson, AZ: University of
Arizona Press, 1999); from L. Chapman, T. Johns, and R.L.A. Mahunnah,
"Saponin-like In Vitro Characteristics of Extracts from Selected Non-nutri-
ent Wild Plant Food Additives Used by Maasai in Meat and Milk Based
Soups," *Ecology of Food and Nutrition*, vol. 36, no. 1 (1997); and from Michael
J. Balick and Paul Alan Cox, *Plants, People and Culture: The Science of
Ethnobotany* (New York: Scientific American Library, 1996).

29. Origins of agriculture from Smith, op. cit. note 27; centers of crop
diversity from N.I. Vavilov, *The Origin and Geography of Cultivated Plants*
(Cambridge, U.K.: Cambridge University Press, 1992); potato species total
from Karl S. Zimmerer, "The Ecogeography of Andean Potatoes," *Bioscience*,
vol. 48, no. 6 (June 1998); Andean crops domesticated from Margery L.
Oldfield, *The Value of Conserving Genetic Resources*, (Washington, DC: U.S.
Department of Interior, National Park Service, 1984); Mario E. Tapia and
Alcides Rosa, "Seed Fairs in the Andes: A Strategy for Local Conservation
of Plant Resources," in David Cooper, René Vellvé, and Henk Hobbelink,
eds., *Growing Diversity: Genetic Resources and Local Food Security* (London:
Intermediate Technology Publications, 1991); and R. Ortega, "Peruvian
In Situ Conservation of Andean Crops," in N. Maxed, B.V. Ford-Lloyd,
and J.G. Hawkes, eds., *Plant Genetic Conservation: The* In Situ *Approach*
(London: Chapman & Hall, 1997); wheat and barley diversity in Ethiopia
from Jack R. Harlan, *Crops and Man* (Madison, WI: American Society of
Agronomy, 1992). Oca is *Oxalis tuberosa*, ullucu is *Ullucus tuberosus*, and
tarwi is *Lupinus mutabilis*.

30. Landrace utility from Michael Loevinsohn and Louise Sperling,
"Joining Dynamic Conservation to Decentralized Genetic Enhancement:
Prospects and Issues," in Michael Loevinsohn and Louise Sperling, eds.,
Using Diversity: Enhancing and Maintaining Genetic Resources On-Farm (New
Delhi: International Development Research Centre (IDRC), 1996); Mexico
example from H. Garrison Wilkes and Susan Wilkes, "The Green
Revolution," *Environment*, vol. 14, no. 8 (1972); Pacific example from Brien
A. Meilleur, "Clones Within Clones: Cosmology and Esthetics and
Polynesian Crop Selection," *Anthropologica*, vol. 40, 1998.

31. India landrace estimate from Swanson et al., op. cit. note 27. Indonesia
details from Michael R. Dove, *Swidden Agriculture in Indonesia: The Subsistence
Strategies of the Kalimantan Kantu'* (Berlin: Walter de Gruyter & Co., 1985);
West Africa details from Paul Richards, *Indigenous Agricultural Revolution*
(London: Hutchinson, 1986). African rice is *Oryza glaberrima*.

32. Developing country seed saving from Mark Wright et al., *The Retention
and Care of Seeds by Small-scale Farmers* (Chatham, U.K.: Natural Resources
Institute (NRI), 1994), and from V. Venkatesan, *Seed Systems in Sub-Saharan
Africa: Issues and Options*, World Bank Discussion Paper 266 (Washington,
DC: World Bank, 1994). Trends in public and private sector agricultural
research from NRC, *Managing Global Genetic Resources: Agricultural Crop
Issues and Policies* (Washington, DC: National Academy Press, 1993).

33. Hybrid seeds from Jack R. Kloppenburg, Jr., *First the Seed: The Political Economy of Plant Biotechnology, 1492–2000* (New York: Cambridge University Press, 1988); from Volker Lehmann, "Patent on Seed Sterility Threatens Seed Saving," *Biotechnology and Development Monitor*, no. 35, June 1998; and from Fowler and Mooney, op. cit. note 3. Greater profitability of hybrids from NRC, op. cit. note 32.

34. Seed-saving illegality from Tracy Clunies-Ross, "Creeping Enclosure: Seed Legislation, Plant Breeders' Rights and Scottish Potatoes," *The Ecologist*, vol. 26, no. 3, May/June 1996, and from John Geadelmann, "Three Main Barriers: Weak Protection for Intellectual Property Rights, Unreasonable Phytosanitary Rules, and Compulsory Varietal Regulation," in David Gisselquist and Jitendra Srivastava, eds., *Easing Barriers to Movement of Plant Varieties for Agricultural Development*, World Bank Discussion Paper 367 (Washington, DC: World Bank, 1997); terminator technology from Lehmann, op. cit. note 33; and from Brian Halweil, "The Emperor's New Crops," *World Watch*, July/August 1999.

35. Examples of inter-species gene transfer from Mann, op. cit. note 1. Transgenic crop data from Clive James, "Global Status and Distribution of Commercial Transgenic Crops in 1997," *Biotechnology and Development Monitor*, no. 35, June 1998; concentration on single gene traits from Halweil, op. cit. note 34.

36. Germplasm contribution of landraces and wild relatives from Don N. Duvick, "Plant Breeding and Biotechnology for Meeting Future Food Needs," in Nurul Islam, ed., *Population and Food in the Early Twenty-First Century: Meeting Future Food Demand of an Increasing Population* (Washington, DC: International Food Policy Research Institute (IFPRI), 1995); from T. Swanson and R. Luxmoore, *Industrial Reliance Upon Biodiversity* (Cambridge, U.K.: World Conservation Monitoring Center, 1997); and from Timothy Swanson, "The Source of Genetic Resource Values and the Reasons for Their Management," in R.E. Evenson, D. Gollin, and V. Santaniello, eds., *Agricultural Values of Plant Genetic Resources* (Wallingford, U.K.: CABI Publishing, 1998).

37. Value of wild relatives from FAO, op. cit. note 8; wine grape history from Jonathan D. Sauer, *Historical Geography of Crop Plants: A Select Roster* (Boca Raton, FL: CRC Press, 1993), and from Rhoades and Nazarea, op. cit. note 12.

38. Contribution of subsistence agriculture to food security and wheat examples from FAO, op. cit. note 8; population estimate for rural poor supported by subsistence agriculture originally calculated by Edward C. Wolf, *Beyond the Green Revolution: New Approaches for Third World Agriculture*, Worldwatch Paper 73 (Washington, DC: Worldwatch Institute, 1986) and updated by Worldwatch Institute using 1997 data from *FAOSTAT Database*, <http://www.apps.fao.org>.

39. Oliver L. Phillips and Brien A. Meilleur, "Usefulness and Economic Potential of the Rare Plants of the United States: A Statistical Survey," *Economic Botany*, vol. 52, no. 1 (1998).

40. H. Garrison Wilkes, "Conservation of Maize Crop Relatives in Guatemala," in C.S. Potter, J.I. Cohen, and D. Janczewski, eds., *Perspectives on Biodiversity: Case Studies of Genetic Resource Conservation and Development* (Washington, DC: AAAS Press, 1993).

41. FAO, op. cit. note 8.

42. Senegal and Gambia example from Tom Osborn, *Participatory Agricultural Extension: Experiences from West Africa* (London: International Institute for Environment and Development, 1995); and from E. Cromwell and S. Wiggins, *Sowing Beyond the State: NGOs and Seed Supply in Developing Countries* (London: Overseas Development Institute, 1993).

43. FAO, op. cit. note 8; Fowler and Mooney, op. cit. note 3; Rural Advancement Foundation International (RAFI), "The Life Industry 1997," *RAFI Communiqué*, November/December 1997; Roundup-ready seeds from Halweil, op. cit. note 34.

44. NRC, op. cit. note 32.

45. India and Bangladesh example from FAO, op. cit. note 8; NRC, op. cit. note 32; Netherlands study from Renée Vellvé, *Saving the Seed: Genetic Diversity and European Agriculture* (London: Earthscan Publications, 1992).

46. Donald L. Plucknett et al., *Gene Banks and the World's Food* (Princeton, NJ: Princeton University Press, 1987); NRC, op. cit. note 32.

47. FAO, op. cit. note 8; estimates of gene bank diversity utilization from Brian D. Wright, *Crop Genetic Resource Policy: Towards A Research Agenda*, Environment and Production Technology Division (EPTD) Discussion Paper 19 (Washington, DC: IFPRI, 1996).

48. Grassy stunt resistance from International Rice Research Institute (IRRI), "Beyond Rice: Wide Crosses Broaden the Gene Pool," <http://www.cgiar.org/irri/Biodiversity/widecrosses.htm>, viewed 9 September 1998.

49. Balick and Cox, op. cit. note 28. Asian snakeroot is *Rauvolfia serpentina* and *R. canescens*.

50. Farnsworth, op. cit. note 7; Balick and Cox, op. cit. note 28; Manjul Bajaj and J.T. Williams, *Healing Forests, Healing People: Report of a Workshop on Medicinal Plants held on 6–8 February, 1995, Calicut, India* (New Delhi: IRDC, 1995); over the counter drug value from Kenton R. Miller and Laura Tangley, *Trees of Life: Saving Tropical Forests and Their Biological Wealth* (Boston, MA: Beacon Press, 1991).

51. The World Health Organization (WHO) figure cited in Balick and Cox, op. cit. note 28; Ayurvedic figures from Bajaj and Williams, op. cit. note 50, and from S.K. Jain and Robert A. DeFilipps, *Medicinal Plants of India, Volume I* (Algonac, MI: Reference Publications, 1991); China figures from Pei-Gen Xiao, "The Chinese Approach to Medicinal Plants," in O. Akerele, V. Heywood and H. Synge, eds., *Conservation of Medicinal Plants* (Cambridge, U.K.: Cambridge University Press, 1991), and from Varro E. Tyler, "Natural Products and Medicine: An Overview," in Michael J. Balick, Elaine Elizabetsky, and Sarah A. Laird, eds., *Medicinal Resources of the Tropical Forest* (New York: Columbia University Press, 1996). Importance of traditional medicine for rural poor from John Lambert, Jitendra Srivastava, and Noel Vietmeyer, *Medicinal Plants: Rescuing a Global Heritage* (Washington, DC: World Bank, 1997), and from Gerard Bodeker and Margaret A. Hughes, "Wound Healing, Traditional Treatments and Research Policy," in H.D.V. Prendergast et al., eds., *Plants for Food and Medicine* (Kew, U.K.: Royal Botanic Gardens, 1998).

52. International medicinal trade details from M. Iqbal, *Trade Restrictions Affecting International Trade in Non-Wood Forest Products*, Non-Wood Forest Products 8 (Rome: FAO, 1995); U.S. herbal market from Jennie Wood Shelton, Michael J. Balick, and Sarah A. Laird, *Medicinal Plants: Can Utilization and Conservation Coexist?* (New York: New York Botanical Garden, 1997).

53. Lambert, Srivastava, and Vietmeyer, op. cit. note 51; A.B. Cunningham, *African Medicinal Plants: Setting Priorities at the Interface Between Conservation and Primary Health Care*, People and Plants Working Paper 1 (Paris: U.N. Educational, Scientific and Cultural Organization (UNESCO), 1993); Panama information from author's fieldnotes, Darien province, Panama, March 1998.

54. A. B. Cunningham and F. T. Mbenkum, *Sustainability of Harvesting Prunus africana Bark in Cameroon: A Medicinal Plant in International Trade*, People and Plants Working Paper 2 (Paris: UNESCO, 1993). African cherry is *Prunus africana*.

55. Paul Hersch-Martínez, "Medicinal Plants and Regional Traders in Mexico: Physiographic Differences and Conservational Challenges," *Economic Botany*, vol. 51, no. 2 (1997). Valerian is *Valeriana* spp.

56. Number of plants assessed from Balick and Cox, op. cit. note 28; undiscovered drug estimates from Robert Mendelsohn and Michael J. Balick, "The Value of Undiscovered Pharmaceuticals in Tropical Forests," *Economic Botany*, vol. 40, no. 2 (1995).

57. Lack of healer apprentices from Balick and Cox, op. cit. note 28; and from Mark Plotkin, *Tales of a Shaman's Apprentice*, (New York: Viking Press, 1994).

58. Percentage of material needs met by plants from the Crucible Group, *People, Plants, and Patents: The Impact of Intellectual Property on Trade, Plant Biodiversity and Rural Society* (Ottawa, Canada: IDRC, 1994); babassu uses from Peter H. May, "Babassu Palm Product Markets," in Plotkin and Famolare, op. cit. note 27; babassu economic importance from Anthony B. Anderson, Peter H. May, and Michael J. Balick, *The Subsidy from Nature: Palm Forests, Peasantry, and Development on an Amazon Frontier* (New York: Columbia University Press, 1991). Babassu is *Attalea speciosa*.

59. Andrew Henderson, Gloria Galeano, and Rodrigo Bernal, *Field Guide to the Palms of the Americas* (Princeton, NJ: Princeton University Press, 1995); Dennis V. Johnson, *Non-wood Forest Products: Tropical Palms*, RAP Publication 10 (Bangkok: FAO, 1997).

60. Ecuador information from Anders S. Barfod and Lars Peter Kvist, "Comparative Ethnobotanical Studies of the Amerindian Groups in Coastal Ecuador," *Biologiske Skrifter*, vol. 46, 1996; Borneo information from Hanne Christensen, "Palms and People in the Bornean Rain Forest," paper presented at Society for Economic Botany Annual Meetings, Aarhus, Denmark, July 1998; India information from Food and Nutrition Division-FAO, "Non-Wood Forest Products and Nutrition," in *Report of the International Expert Consultation on Non-Wood Forest Products*, Non-Wood Forest Products Series 3 (Rome: FAO, 1995).

61. Timber tree total calculated from trade statistics of the International Tropical Timber Organization, <http://www.itto.or.jp/timber_situation/timber1997/tables/appendix.html>, viewed 7 September 1998; essential oil total from J.J.W. Coppen, *Flavours and Fragrances of Plant Origin*, Non-Wood Forest Products Series 1 (Rome: FAO, 1995); gum and latex total from J.J.W. Coppen, *Gums, Resins and Latexes of Plant Origin*, Non-Wood Forest Products Series 6, (Rome: FAO, 1995); dye total from C.L. Green, *Natural Colourants and Dyestuffs*, Non-Wood Forest Products Series 4 (Rome: FAO, 1995).

62. Sara Oldfield, ed., *Cactus and Succulent Plants: Status Survey and Conservation Action Plan* (Gland, Switzerland: IUCN, 1997).

63. Terry Sunderland, "The Rattan Palms of Central Africa and Their Economic Importance," paper presented at Society for Economic Botany Annual Meetings, Aarhus, Denmark, July 1998. Regional status of rattan resources from International Network for Bamboo and Rattan (INBAR), "Bamboo/Rattan Worldwide" Reports, *INBAR Newsletter*, vol. 4 (no date) and vol. 5 (1997).

64. Monte Basgall, "Logging is Creating Mahogany Deserts," Duke Research 1996–1997, <http://www.dukeresearch.duke.edu/dukeresearch/dr97/mahogany.htm>, viewed 22 July 1999; Tod Riggio, "Limited supplies keep mahogany prices high," <http://www.woodshopnews.com/stories/mahog/>, viewed 22 July 1999. American mahogany is *Swietenia macrophylla* and *Swietenia mahogani*.

65. Wild tomato valuation from Hugh H. Iltis, "Serendipity in the Exploration of Biodiversity: What Good Are Weedy Tomatoes?" in Wilson, ed., op. cit. note 7.

66. Nature's services detailed in Gretchen Daily, ed., *Nature's Services: Societal Dependence on Natural Ecosystems* (Washington, DC: Island Press, 1997); long-nosed bats from Stephen L. Buchmann and Gary Paul Nabhan, *The Forgotten Pollinators* (Washington, DC: Island Press, 1996).

67. Polyculture figures and home garden diversity from Lori Ann Thrupp, *Cultivating Diversity: Agrobiodiversity and Food Security* (Washington, DC: World Resources Institute (WRI), 1998).

68. Ecological benefits of polycultures from Richards, op. cit. note 31; Miguel A. Altieri, *Agroecology: The Science of Sustainable Agriculture* (Boulder, CO: Westview Press, 1995, 2nd ed.); and from Steven R. Gliessman, ed., *Agroecology: Researching the Basis of Sustainable Agriculture* (Berlin: Springer-Verlag, 1990).

69. Venezuela example from L. Sarmiento, M. Monasterio, and M. Montilla, "Ecological Bases, Sustainability and Current Trends in Traditional Agriculture in the Venezuelan High Andes," *Mountain Research and Development*, vol.13, no. 2 (1993).

70. Lori Ann Thrupp, Susanna Hecht, and John Browder, *The Diversity and Dynamics of Shifting Cultivation: Myths, Realities and Policy Implications* (Washington, DC: WRI, 1997).

71. Altieri, op. cit. note 68; Sarmiento, Monasterio, and Montilla, op. cit. note 69.

72. Biological dynamics of fragmented habitats from William S. Alverson, Walter Kuhlmann, and Donald M. Waller, *Wild Forests: Conservation Biology and Public Policy* (Washington, DC: Island Press, 1993); island and area effects were first synthesized by Robert H. MacArthur and Edward O. Wilson, *The Theory of Island Biogeography* (Princeton, NJ: Princeton University Press, 1967).

73. Río Palenque details from C.H. Dodson and A.H. Gentry, "Biological Extinction in Western Ecuador," *Annals of the Missouri Botanical Garden*, vol. 78, no. 2 (1991); and from Gentry, op. cit. note 27.

74. Richard B. Primack, *Essentials of Conservation Biology* (Sunderland, MA: Sinauer Associates, 1993).

75. Sarmiento, Monasterio, and Montilla, op. cit. note 69; M. Molinillo and M. Monasterio, "Pastoralism in Paramo Environments: Practices, Forage and Impact on Vegetation in the Cordillera de Mérida, Venezuela," *Mountain Research and Development*, vol. 17, no. 3 (1997); R. Vicente Casanova, ed., *La*

Agricultura Campesina en la Subregión de Mucuchíes (Mérida, Venezuela: Universidad de los Andes, Consejo de Publicaciones, 1998).

76. Sarmiento et al., op. cit. note 69; Casanova, op. cit. note 75.

77. Health risks and limited regulatory oversight of pesticides are documented in Theo Colbourn, Dianne Dumanoski, and John Peterson Myers, *Our Stolen Future* (New York: Dutton Books, 1996), and in John Wargo, *Our Childrens' Toxic Legacy: How Science and Law Fail to Protect Us from Pesticides* (New Haven, CT: Yale University Press, 1997); glyphosate example from Fred Pearce and Deborah Mackenzie, "It's Raining Pesticides," *New Scientist*, 3 April 1999.

78. Switzerland example from Pearce and Mackenzie, op. cit. note 77; developing-country pesticide problems from Lori Ann Thrupp, *Bittersweet Harvests for Global Supermarkets: Challenges in Latin America's Export Boom* (Washington, DC: WRI, 1995).

79. Population figures from U.N. Population Division, op. cit. note 6.

80. Harlan, op. cit. note 29.

81. Botanical garden conservation statistics from Vernon H. Heywood, "Broadening the Basis of Plant Resource Conservation," in J. Perry Gustafson, R. Appels, and P. Raven, eds., *Gene Conservation and Exploitation*, (New York: Plenum Press, 1993), and from Botanic Gardens Conservation International, "Introduction: Botanic Gardens & Conservation," <http://www.rbgkew.org.uk/BGCI/babrief.htm>, viewed 17 June 1998.

82. Plucknett et al., op. cit. note 46; Botanical Gardens Conservation International, op. cit. note 81; information on CPC from Jane Bosveld, "A Green and Flowering Place," *Scientific American Explorations*, winter 1999.

83. Fowler and Mooney, op. cit. note 3; J.G. Hawkes, *The Diversity of Crop Plants* (Cambridge, MA: Harvard University Press, 1983); Millennium Seed Bank information from R.D. Smith, S.H. Linington, and G.E. Wechsberg, "The Millennium Seed Bank, The Convention on Biological Diversity and the Dry Tropics," in H.D.V. Prendergast et al., op. cit. note 51.

84. Hawkes op. cit. note 83; Mooney, op. cit. note 3.

85. FAO, op. cit. note 8; Plucknett et al., op. cit. note 46; Wright, op. cit. note 47; landrace coverage in gene banks is questioned by Fowler and Mooney, op. cit. note 3, and by Pablo Eyzaguirre and Masa Iwanaga, "Farmers' Contribution to Maintaining Genetic Diversity in Crops, and Its Role Within the Total Genetic Resources System," in P. Eyzaguirre and M. Iwanaga, eds., *Participatory Plant Breeding: Proceedings of a Workshop on Participatory Plant Breeding, 26–29 July 1995, Wageningen, The Netherlands* (Rome: International Plant Genetic Resources Institute (IPGRI), 1996).

86. Total annual cost for accession maintenance calculated from figure of $50 per accession from Stephen B. Brush, "Valuing Crop Genetic Resources," *Journal of Environment & Development*, vol. 5, no. 4 (December 1996); Plucknett et al., op. cit. note 46; FAO, op. cit. note 8; maize regeneration project from Major Goodman, "Research Policies Thwart Potential Payoff of Exotic Germplasm," *Diversity*, vol. 14, nos. 3/4 (1998).

87. FAO, op. cit. note 8; Plucknett et al., op. cit. note 46; NRC, op. cit. note 32.

88. Indian palm extirpation from Dennis Johnson, ed., *Palms: Their Conservation and Sustained Utilization*, Status Survey and Conservation Action Plan (Gland, Switzerland: IUCN, 1996).

89. Rwanda example from Louise Sperling, "Results, Methods, and Institutional Issues in Participatory Selection: The Case of Beans in Rwanda," in Eyzaguirre and Iwanaga, op. cit. note 85; West Africa example from Paul Richards and Guido Ruivenkamp, *Seeds and Survival: Crop Genetic Resources in War and Reconstruction in Africa* (Rome: IPGRI, 1997).

90. Gary Paul Nabhan, *Enduring Seeds: Native American Agriculture and Wild Plant Conservation* (San Francisco, CA: North Point Press, 1989).

91. Global protected area total from IUCN, <http://www.iucn.org/info_ and_news/index.html>, viewed 20 September 1998; Mt. Kinabalu from Heywood and Davis, op. cit. note 13; Indian citrus reserve from FAO, op. cit. note 8; Ammiad reserve from N. Maxted, B.V. Ford-Lloyd, and J.G. Hawkes, eds., op. cit. note 29.

92. Limitations of protected areas from Carel van Schaik, John Terborgh, and Barbara L. Dugelby, "The Silent Crisis: The State of Rain Forest Nature Preserves," in Randall Kramer, Carel van Schaik, and Julie Johnson, eds., *Last Stand: Protected Areas and the Defense of Tropical Biodiversity* (New York: Oxford University Press, 1997).

93. P.S. Ramakrishnan, "Conserving the Sacred: From Species to Landscapes," *Nature & Resources*, vol. 32, no. 1 (1996); West Africa details from Aiah R. Lebbie and Raymond P. Guries, "Ethnobotanical Value and Conservation of Sacred Groves of the Kpaa Mende in Sierra Leone," *Economic Botany*, vol. 49, no. 3 (1995); Pacific Islander information from Balick and Cox, op. cit. note 28; status of rarest U.S. plants from Brien A. Meilleur, Center for Plant Conservation, letter to author, 30 June 1999.

94. India information from Mark Poffenberger, Betsy McGean, and Arvind Khare, "Communities Sustaining India's Forests in the Twenty-first Century," in Mark Poffenberger and Betsy McGean, eds., *Village Voices, Forest Choices: Joint Forest Management in India* (New Delhi: Oxford University Press, 1996); and from Payal Sampat, "India's Choice," *World Watch*, July/August 1998.

95. Belize information from Balick and Cox, op. cit. note 28; from Rosita Arvigo, *Sastun: My Apprenticeship with a Maya Healer* (New York: Harper Collins, 1994); and from Michael J. Balick, Rosita Arvigo, Gregory Shropshire, and Robert Mendelsohn, "Ethnopharmacological Studies and Biological Conservation in Belize," in Balick, Elizabetsky, and Laird, eds., op. cit. note 51.

96. Convention on Biodiversity Secretariat, <http://www.biodiv.org>, viewed 22 April 1998; IUCN, *The Global Environment Facility-From Rio to New Delhi: A Guide for NGOs* (Gland, Switzerland, and Cambridge, U.K.: 1997); Anatole F. Krattiger et al., eds., *Widening Perspectives on Biodiversity* (Gland, Switzerland: IUCN and International Academy of the Environment, 1994). Table 6 was previously published in Worldwatch Paper 141.

97. Amounts pledged to GEF for biodiversity conservation calculated from Hilary F. French, *Partnership for the Planet: An Environmental Agenda for the United Nations*, Worldwatch Paper 126 (Washington, DC: Worldwatch Institute, July 1995); independent assessments of GEF performance from IUCN, op. cit. note 96; and from Rob Edwards and Sanjay Kumar, "Dust to Dust," *New Scientist*, 6 June 1998.

98. Michael R. Dove, "A Revisionist View of Tropical Deforestation and Development," *Environmental Conservation*, vol. 20, no. 1 (1993).

99. Subsistence land use patterns and land use changes in Kalimantan from Christine Padoch and Nancy Lee Peluso, eds., *Borneo in Transition: People, Forests, Conservation and Development* (Oxford, U.K.: Oxford University Press, 1996); political conflicts and growth of plantation sector from Ricardo Carrere and Larry Lohmann, *Pulping the South: Industrial Tree Plantations and the World Paper Economy* (London: Zed Books, 1996) and from Koesnadi Wirasapoetra, "Palm Oil in the Land of the Pasir Community: Promise, and Control Over the Land," *Terompet*, no. 11/IV, 1997.

100. Details on 1997–98 fires from "Forest Fires in East Kalimantan Bring Famine and Dispossession," Down to Earth news release, 11 June 1998, from Andrew P. Vayda, *Finding Causes of the 1997–98 Indonesian Forest Fires: Problems and Possibilities* (Jakarta: WWF Indonesia Forest Fires Project, 1999), and from Emily Harwell, "Indonesian Inferno" and "Unnatural Disaster," *Natural History*, July-August 1999.

101. A.P. Vovides and C.G. Iglesias, "An Integrated Conservation Strategy for the Cycad Dioon edule Lindl," *Biodiversity and Conservation*, vol. 3, 1994; Andrew P. Vovides, "Propagation of Mexican Cycads by Peasant Nurseries," *Species*, December 1997.

102. Vovides and Iglesias, op. cit. note 101; Vovides, op. cit. note 101; examples of requirements for sustainability from Oliver Phillips, "Using and Conserving the Rainforest," *Conservation Biology*, vol. 7, no. 1 (March 1993); from Jason W. Clay, *Generating Income and Conserving Resources: 20 Lessons* *from the Field* (Washington, DC: WWF-U.S., 1996); and from Charles M. Peters, *Sustainable Harvest of Non-timber Plant Resources in Tropical Moist Forest: An Ecological Primer* (Washington, DC: United States Agency for International Development (USAID), Biodiversity Support Program, 1994).

103. Vovides and Iglesias, op. cit. note 101; Vovides, op. cit. note 101; "CITES and the Royal Botanic Gardens Kew," <http://www.rbgkew.org.uk/ ksheets/cites.html>, viewed 17 June 1998.

104. Environmental certification programs for timber from Janet N. Abramovitz, "Sustaining the World's Forests," in Lester R. Brown, Christopher Flavin, and Hilary French, *State of the World 1998* (New York: W.W. Norton & Company, 1998); total hectares under certification from FAO, *State of the World's Forests 1999* (Rome: 1999).

105. International Technical Conference on Plant Genetic Resources, *Global Plan of Action for the Conservation and Sustainable Utilization of Plant Genetic Resources for Food and Agriculture* (Rome: FAO, 1996); budget figure and financing details from FAO, "Financing the Implementation of the Global Plan of Action on Plant Genetic Resources," presented at Eighth Regular Session of the Commission on Genetic Resources for Food and Agriculture, Rome, Italy, April 1999.

106. FAO, "Facilitating and Monitoring the Implementation of the Global Plan of Action on Plant Genetic Resources," presented at the Eighth Regular Session of the Commission on Genetic Resources for Food and Agriculture, Rome, Italy, April 1999.

107. Karl S. Zimmerer, *Changing Fortunes: Biodiversity and Peasant Livelihood in the Peruvian Andes* (Berkeley, CA: University of California Press, 1996); Hopi example from Nabhan, op. cit. note 90; Mende example from NRC, op. cit. note 27.

108. Zimbabwe example from Seema van Oosterhout, "What Does *In Situ* Conservation Mean in the Life of a Small-Scale Farmer? Examples from Zimbabwe's Communal Areas," in Loevinsohn and Sperling, eds., op. cit. note 30; other information from Eyzaguirre and Iwanaga, eds., op. cit. note 85.

109. J.R. Whitcombe and A. Joshi, "The Impact of Farmer Participatory Research on Biodiversity of Crops," in Loevinsohn and Sperling, op. cit. note 30; John Whitcombe and Arun Joshi, "Farmer Participatory Approaches for Varietal Breeding and Selection and Linkages to the Formal Seed Sector," in Eyzaguirre and Iwanaga, op. cit. note 85; K.W. Riley, "Decentralized Breeding and Selection: Tool to Link Diversity and Development," in Loevinsohn and Sperling, op. cit. note 30.

110. Lawrence Busch et al., *Making Nature, Shaping Culture: Plant Biodiversity in Global Context* (Lincoln, NE: University of Nebraska, 1995); Seed Saver's

Exchange, "Welcome to the Seed Saver's Exchange," members' informational brochure, 1992.

111. Busch et al., op. cit. note 110; diabetes and native foods from Gary Paul Nabhan, *Cultures of Habitat* (Washington, DC: Counterpoint Press, 1997); and from John Tuxill and Gary Paul Nabhan, *Plants and Protected Areas: A Guide to In Situ Management* (Cheltenham, U.K.: Stanley Thornes, 1998).

112. Centralization of modern agriculture from James C. Scott, *Seeing Like A State: How Certain Schemes to Improve the Human Condition Have Failed* (New Haven, CT: Yale University Press, 1998); Swanson, Pearce, and Cervigni, op. cit. note 27.

113. Eyzaguirre and Iwanaga, eds., op. cit. note 85; G.M. Listman, G. Srinivasan, S. Taba, and M. Smale, "Mexican and CIMMYT Researchers 'Transform Diversity' of Highland Maize to Benefit Farmers with High-Yielding Seed," *Diversity*, vol. 12, no. 2 (1996).

114. Southern African subsidies from FAO, op. cit. note 8; and Swanson, Pearce, and Cervigni, op. cit. note 27.

115. Thrupp, op. cit. note 78.

116. Iowa information from Lori Ann Thrupp, ed., *New Partnerships for Sustainable Agriculture* (Washington, DC: WRI, 1996); and from Brian DeVore, "Changing the Land's Complexion," *The Land Stewardship Letter*, vol. 16, no. 6 (December 1998).

117. Rice pedigree information from D. Gollin and R.E. Evenson, "Breeding Values of Rice Genetic Resources," in Evenson, Gollin, and Santaniello, op. cit. note 36; and from IRRI, "Delivering Diversity to the Field," <http://www.cgiar.org/irri/Biodiversity/pedigreediversity.htm>, viewed 4 September 1998; wheat information from M. Smale, "Indicators of Varietal Diversity in Bread Wheat Grown in Developing Countries," in Evenson, Gollin, and Santaniello, op. cit. note 36.

118. Organic agriculture trends from Thrupp, op. cit. note 67; role of farmers' markets from John Tuxill, "Why is the Market So Diverse? Crop Varieties in the Madison, Wisconsin, Farmers' Market," unpublished manuscript, December 1993; slow-food movement from *The Snail: A Slow Food Newsletter*, Bra, Italy, Issue 1, 1999.

119. Mark Lipson, *Searching for the "O-Word": An Analysis of the U.S.D.A. Current Record Information System (CRIS) for Pertinence to Organic Farming* (Santa Cruz, CA: Organic Farming Research Foundation, 1997).

120. Brush, op. cit. note 86.

121. Stephen B. Brush, "Is Common Heritage Outmoded?" in Stephen B.

Brush and Doreen Stabinsky, eds., *Valuing Local Knowledge: Indigenous People and Intellectual Property Rights* (Washington, DC: Island Press, 1996).

122. Mooney, op. cit. note 3; Tuxill and Nabhan, op. cit. note 111; Brush, op. cit. note 86; Brush, op. cit. note 121.

123. Darrell A. Posey, *Traditional Resource Rights: International Instruments for Protection and Compensation for Indigenous Peoples and Local Communities* (Gland, Switzerland: IUCN, 1996).

124. InBio details from Michel P. Pimbert and Jules Pretty, *Parks, People and Professionals: Putting "Participation" Into Protected Area Management*, UNRISD Discussion Paper #57 (Geneva: U.N. Research Institute for Social Development, 1995); and from Anthony Artuso, "Capturing the Chemical Value of Biodiversity: Economic Perspectives and Policy Prescriptions," in Francesca Grifo and Joshua Rosenthal, *Biodiversity and Human Health* (Washington, DC: Island Press, 1997).

125. Drug development cost estimates from Shelton, Balick, and Laird, op. cit. note 52; bioprospecting details from Pimbert and Pretty, op. cit. note 124, and from RAFI, "Biopiracy Update: The Inequitable Sharing of Benefits," *RAFI Communiqué*, September/October 1997; traditional resource rights from Posey, op. cit. note 123.

126. Crucible Group, op. cit. note 58; Swanson, Pearce, and Cervigni, op. cit. note 27; Bees Butler and Robin Pistorius, "How Farmers' Rights Can Be Used to Adapt Plant Breeders' Rights," *Biotechnology and Development Monitor*, no. 28, September 1996; Henry L. Shands, "Access: Bartering and Brokering Genetic Resources," in June F. MacDonald, ed., *Genes for the Future: Discovery, Ownership, Access*, NABC Report #7 (Ithaca, NY: National Agricultural Biotechnology Council, 1995); RAFI, "Repeat the Term! Report on FAO's Gene Commission in Rome June 8–12, 1998," *RAFI Occasional Paper Series*, vol. 5, no. 2 (July 1998).

127. Butler and Pistorius, op. cit. note 126; Crucible Group, op. cit. note 58; José Luis Solleiro, "Intellectual Property Rights: Key to Access or Entry Barrier for Developing Countries," in MacDonald, ed., op. cit. note 126.

128. RAFI, "The Australian PBR Scandal," *RAFI Communiqué*, January/February 1998; Anon., "Foreign Invasion," *Down to Earth*, 28 February 1998; Rob Edwards and Ian Anderson, "Seeds of Wrath," *New Scientist*, 14 February 1999.

129. Meilleur, op. cit. note 17.

Worldwatch Papers

No. of Copies

Worldwatch Papers by John Tuxill

_____**Total copies (transfer number to order form on next page)**

PUBLICATION ORDER FORM

NOTE: Many Worldwatch publications can be downloaded as PDF files from our website at **www.worldwatch.org**. Orders for printed publications can also be placed on the web.

_____ *State of the World:* $13.95
The annual book used by journalists, activists, scholars, and policymakers worldwide to get a clear picture of the environmental problems we face.

_____ **State of the World Library: $30.00 (international subscribers $45)**
Receive *State of the World* and all five Worldwatch Papers as they are released during the calendar year.

_____ *Vital Signs:* $13.00
The book of trends that are shaping our future in easy-to-read graph and table format, with a brief commentary on each trend.

_____ **WORLD WATCH magazine subscription: $20.00 (international airmail $35.00)**
Stay abreast of global environmental trends and issues with our award-winning, eminently readable bimonthly magazine.

_____ **Worldwatch Database Disk Subscription: $89.00**
Contains global agricultural, energy, economic, environmental, social, and military indicators from all current Worldwatch publications. Includes a mid-year update, and *Vital Signs* and *State of the World* as they are published. Disk contains Microsoft Excel spreadsheets 5.0/95 (*.xls) for Windows.
Check one: _____ **PC** _____ **Mac**

_____ **Worldwatch Papers—See list on previous page**
Single copy: $5.00
2–5: $4.00 ea. • 6–20: $3.00 ea. • 21 or more: $2.00 ea.

$4.00* Shipping and Handling *($8.00 outside North America)*
*minimum charge for S&H; call (800) 555-2028 for bulk order S&H

_____ **TOTAL** (U.S. dollars only)

Make check payable to: Worldwatch Institute, 1776 Massachusetts Ave., NW, Washington, DC 20036-1904 USA

Enclosed is my check or purchase order for U.S. $_____

☐ AMEX   ☐ VISA   ☐ MasterCard _____
                                     Card Number   Expiration Date

signature _____

name _____ **daytime phone #**

address _____

city _____ **state**   **zip/country**

phone: (800) 555-2028   fax: (202) 296-7365   e-mail: wwpub@worldwatch.org
website: www.worldwatch.org

Wish to make a tax-deductible contribution? Contact Worldwatch to find out how your donation can help advance our work.